Casting Demons Into Swine

A Novel

R.J. Erskine

Mil gracias y reconocimientos to **Mark Siragusa** for photography, copyright © 2015 by Mark T. Siragusa, All rights reserved
http://www.mtsphoto.com/galleries

Admiration and acknowledgement to **Chris Herron** for graphic and cover design, copyright © 2016 by Chris Herron.
https://www.chrisherrondesign.com/

All rights reserved. Special thanks to **Hannah Gregus** for seasonal graphic art.

Finally, my gratitude to Pam and Paul, Kym dear, Sebastian Video, MOE, and above all, Miss Constance for creative insight and critique.

Erskine, Ronald,
    Casting demons into swine / RJ Erskine.
    "A Stray Voltage Press book"
    ISBN 978-0-9971873-2-8

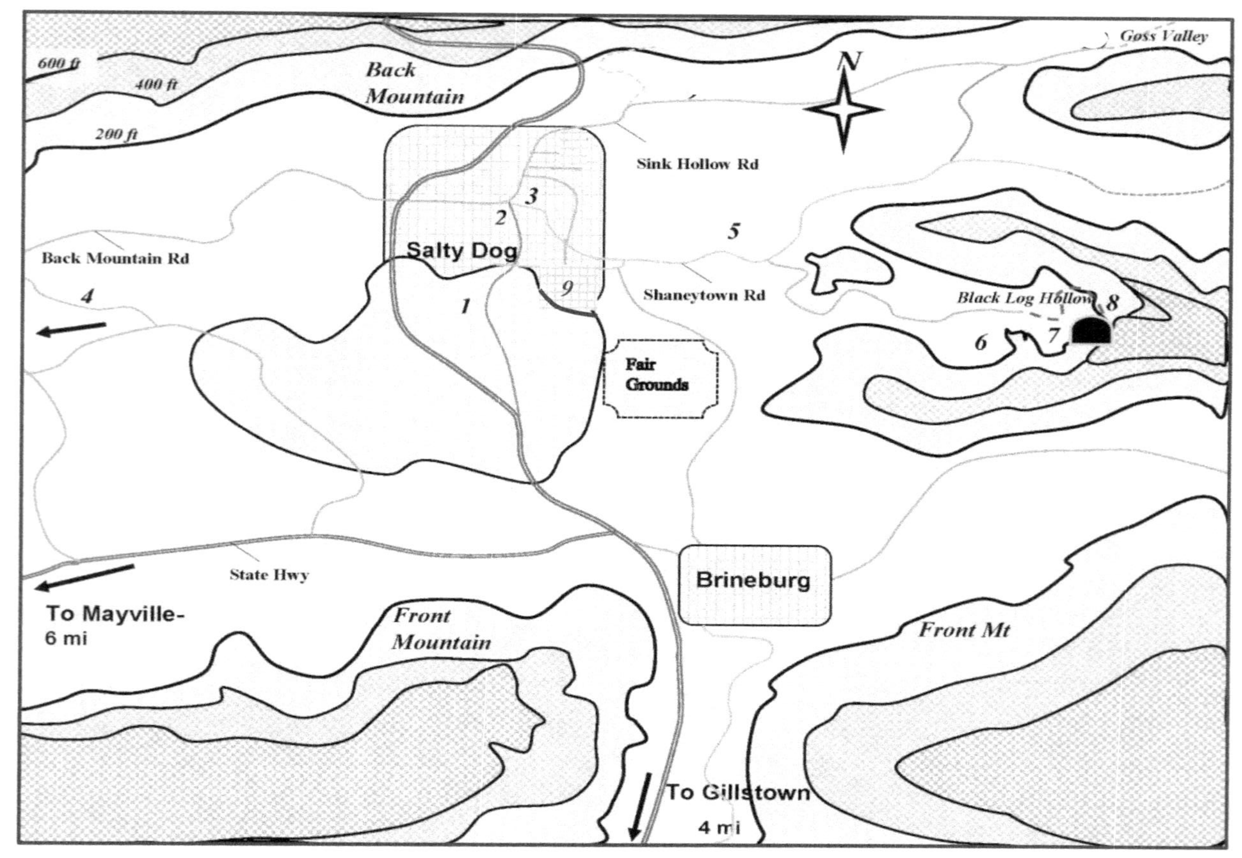

600 ft
400 ft
200 ft
Back Mountain
Back Mountain Rd
4
Sink Hollow Rd
3
2
9
1
5
Salty Dog
Shaneytown Rd
Fair Grounds
6
7
8
Black Log Hollow
Cross Valley
N
State Hwy
To Mayville-
6 mi
Front Mountain
Brineburg
Front Mt
To Gillstown
4 mi

LEGEND

1- Dr. Malcolm Cromarty House

2- Dead Bear Bar

3- Salty Dog Hose Company No. 1

4- John Y. Peachey and Nahum Y. Peachey Farms

5- Dale Calhoun Farm

6- Jonah P. Yoder Farm

7- McDougall Homestead

8- Dark Spring Cave

9- Cap's Market

Prologue

David Alexander MacDougall pushed his way through the brush and rocks on the steep slope, hoping to seize his own stake in the ancestral lands of another race. He sought to leave the festering cities behind, full of refugees from overseas, carrying their baggage of disease, poverty, and misery. But he also needed to get beyond the reach of the law. Reaching the summit the summit of a level-topped ridge, his grim grey eyes looked upon ranks of distant ridges, their crests undulating along an endless horizon, reminding him of the swell of the ocean he had crossed five years before. Thinking of the ship that had carried him to this continent, he felt free from his past once again.

He had paid dearly for this moment. Five long years as an indentured servant to a blacksmith who toiled long hours and forced David to work even longer. Five long years, where each day was just another weary milestone of his debt for passage from Belfast to the colonies. Each hour at the forge wasn't marked by a clock, but by drops of sooty sweat, crushed fingernails, and arm and back muscles seared with pain, as he twisted iron to obey the will of his hammer. He had survived the pox, a burn that blistered and scarred his cheek, and the loss of two fingers on his left hand.

Unable to marry without his master's permission, David MacDougall secretly courted the blacksmith's daughter, waiting for the day of freedom to offer his hand in marriage. His one purpose in life, dawn to dusk, was to endure through an eternity of five years. That day finally arrived and David, no longer bound, posed the question to his master. The blacksmith laughed at the proposal, replying that arrangements had already been made with another suitor of better means, adding derisively, who also had all his fingers. His overseer then offered a consolation prize, David was welcome to stay five additional days on his tattered straw mattress in a corner of the floor of the blacksmith shop, and all the stale bread and bone broth he could eat, one day for each year of his service. Of course, he was expected to work in the shop for that gift but would be paid a copper penny for each day worked. But he must be gone in five days—another work-slave was arriving from overseas soon, and the blacksmith had no need for a paid laborer.

David's bitterness spread a canker within his mind, connecting with deep ancestral roots from the brutal borders of Scotland. He simmered until the end of the fifth day. After bidding farewell to the blacksmith and his family, seemingly to find his fortune, David doubled back after dark to find his master already in his house and the shop abandoned for the night.

He approached the back door of the shop. As expected, it was unlocked, the blacksmith had overlooked his new task. David thrust what tools he could lay his hands on into a leather sack and grabbed the musket and powder from a hiding place in the loft above the forge. He entered the stable, bridled his master's favorite chestnut gelding and carefully lashed the tools onto the saddle. Save but a four-pound hammer, which he grasped as he crossed the dim alley towards the back of the house.

David lifted the latch of the door to find the kitchen empty, a fire burning in the hearth. The blacksmith and his family were noisily consuming stew, bread and ale in the dining room. He surveyed the kitchen and scooped handfuls of lard from a tin, rubbed the grease onto a couple of towels and wrapped the towels around the legs of a bench by the wall. Deftly plucking a flaming log from the fire with tongs, he laid it under the bench and slid behind the door.

Within minutes, flames fed by the rags spouted into the bottom of the bench, scorching the pitch pine boards. Smoke rolled across the ceiling, changing the raucous conversation in the dining room into shouts and crashes of tumbled chairs. The blacksmith's wife appeared first. David shoved her aside, his strong arms sending her careening onto the floor. The blacksmith's stout body filled the doorway, trying to make sense of the fire, his wife crawling on the floor and the presence of David all in one take. Without a sound, David whipped a backhand stroke of the hammer into the smith's temple, his body collapsing under the blow like a deflated balloon. While his former lover and her sisters stood transfixed from the chaos of David's handiwork, he lifted a loaf of bread and a cheese from a rack, strode back to the stable and trotted away from the 'City of Brotherly Love' on the murdered man's horse.

He rode north and west along rutted coach roads that dissolved into meager wagon tracks along the river. Leaving the river, he pressed along narrow footpaths that hugged the banks of tributaries and creeks, cutting through notches and gaps in the ever-rising ridges, where settlements and cabins of white men were scarce. At last, he came upon a deep recess in a dark forest of hemlock, where there was no longer a path, just the cut from a gurgling stream winding through the roots of the trees.

Leaving the horse and his pack by the burn, he climbed to gain the view he now beheld. Wanting to unpack and settle before sunset, he descended and came across a dark pool of water fed by a steady trickle over a ledge. Tracking the source of the flow, he was surprised to see a large cleft in the face of the rock above the ledge. Approaching with caution, his musket ready, the twenty-foot-wide fissure proved to be the entrance to a large chamber, which seemed to go deep into the mountain. Later, after he had set camp, he returned with a torch and entered the cave.

Trained eyes of a blacksmith pegged red hued streaks within the walls of the cave—iron ore of high quality. David quickly surmised that limestone and wood for charcoal, needed for a furnace to smelt the ore into iron, were in good supply. This could be a source of iron for not only his craft, but eventually, blacksmiths for miles around.

For many years after, clan MacDougall flourished from the ore mined near Dark Spring Cave as the demand for iron grew around them. With the growing population came an appetite for lumber, and railroads carried the plunder from the forests near and far. Rails of steel paved a fortune of gold for the MacDougalls that lasted well into the century after David had stumbled on his discovery.

With time, as the supply of lumber and ore waned, petty disputes between neighbors over poorly surveyed boundaries erupted into ceaseless feuds. Among them, the Calhouns and MacDougalls emerged as the most rabid of enemies. Bodies of kinfolk who met premature and violent ends rested throughout the rocky hollows and ridges.

Yet, fate intervened to usher in a new style of entrepreneurship in the valley—prohibition—bringing a second golden age for the hill clans. The natural harmony of wood, pure water, and good limestone soil to grow corn melded together into a new trade, distilling. Family wealth soared with the public's dislike for forced sobriety and the windfall was shared liberally with politicians and the law. But the Twenty-First Amendment shut down the cash tap and the Great Depression cast the MacDougalls into a financial abyss.

To maintain payments of taxes and other debts, the family was forced to sell off much of their best tillable land. Their German-speaking neighbors, aware of the MacDougall's dire straits, were always there to purchase the parcels, often at a devalued price. Angered by their neighbor's willingness to grasp cheap land, the MacDougalls reciprocated by resorting to another ancestral skill, rustling livestock. Stories of missing cattle and sheep pervaded the lower end of the valley, but there was little desire to confront the thieves or involve the law. As time passed, clan MacDougall grew more reclusive. Prudent people shunned Black Log Hollow at the lower end of the valley, where the hills squeezed together like massive tongs. Even hunters caught in the thrill of buck season avoided this disquieting place. As for the MacDougalls, they relished their isolation and cultivated what they did best, hate. Hatred for the Calhouns and most of all, the 'dutchmen'.

SUMMER

CHAPTER I

Dear Malcolm,
Your lack of caring is unbearable. I tried to make this work, but your career is more important than our marriage. Perhaps if I ate grass and had black and white spots I might be more worthy of your attention. I can't live here anymore; it's hell, not home. I'm sick and tired of being walled up in the shadow of these gloomy hills where no one has a private life in this damn valley and everyone knows everyone else's business. This place gives me the creeps and I can't stay another day. You're fooling yourself to think you'll ever fit in here, at least I knew we don't belong. I'm going back home to reset my life; my father found a position for me at a local bank. He always said I never should have followed you out here, you're in a dead-end job and will never be successful. After our argument last night, I'm convinced he was right. I didn't want to leave without saying goodbye but feel it's better for both of us. I'll find a lawyer when I get home and contact you about a settlement for the house. I took the truck to haul my furniture. Lucille's all yours. I'll miss Precious and to some extent, you. But I've had enough, hope you get your act together someday.

Carrie

CHAPTER II

Malcolm Cromarty crash-landed into reality with ringing in his ears. His fingers traced the coiled wire and grasped the handset. Struggling to gain his voice, he grumbled, "Hello, Dr. Cromarty."

A timid adolescent male voice responded, "John Y. Peachey has a milk fever."

Malcolm processed the message; task was revealed but he lacked location. "John Y. on Honey Creek Road?" Malcolm eased two paws off his chest.

"No, John Y. Peachey on Rocky Creek Road. She's down up in the pasture."

Grimacing as a wet tongue licked his ear, Malcolm raised himself onto the edge of his bed. "Okay, I'll be there." He replaced the handset and squinted against the glare as he pulled the chain on the table lamp. It was barely five o'clock. He thought. Nice start to a day. 'Down up' in the pasture? What does that mean? Precious leaned against his legs, wagging her tail. Malcolm wiped his hand on the hair between her ears, "Think you're pretty damn clever, don't you?"

Precious trotted from the room. Malcolm remained behind, preferring to continue his rendition of a seated mannequin. Precious reappeared in the doorway, spinning her head in a tight circle, "Woof!"

"Relax! For God's sake." Malcolm gathered himself onto his feet. "Stop pestering me and maybe I'll take you for a ride this morning." Precious bolted away, the hallway echoed with the sound of her nails ticking on the hard wood floor.

"Woof!"

"What's with you? Jeez, you can be such a pain in the ass!" Malcolm shuffled down the hallway to find her waiting by the back door. He pulled the door open. Precious rose onto her rear feet, slapped the latch of the screen door and plunged into the murky grey light of the back yard.

Nearly sleepwalking, Malcolm tracked his hand along the kitchen wall and flicked a switch. Fluorescent light splattered the space with a milky glow. Grabbing a can of instant coffee, he dug out a spoon and mug from a haphazard collection of utensils and cookware in the sink. Peering at a stain in the bottom, he rinsed the mug and popped open the coffee can lid, managing to scatter freeze dried granules over the entirety of the unwashed dishes. "Piss on it, I'll live without it."

Returning to his bedroom, he rummaged for jeans and a tee shirt in a pile of laundry, ambled to the bathroom, emptied his bladder, brushed his teeth, and recoiled from a splash of cold water on his face. His eyes took stock of the brown-eyed zombie with tangled sandy hair staring at him from the mirror. Returning to the back door, he called out, "Let's go, screwball!"

Precious careened off Malcolm's calves as she sped into the house. "Yeah, yeah, just relax." She continued her headlong charge as he opened the front door, sprinted along the walk and planted her front paws on the handle of the driver's door of an International Scout. "Go ahead. Let's see you open that one."

Light dew dampened Malcolm's palm as he yanked the door handle. Precious wasted no time clambering onto the seat. Scattered in the branches of the trees around the house, tentative songbirds were the only audible evidence of other life on the planet. A breeze sifted from the pastel canvas of the eastern sky, teasing the hair on Malcolm's arms. He sighed. If only it would stay this cool the rest of the day.

Malcolm opened the back hatch and released vapors of stale bovine, seasoned with road dust, from stained cardboard boxes. Sorting through a collection of solutions, powders, sprays, and disinfectants, his hands pulled out a couple bottles of calcium solution. He leaned further through the hatch and disentangled a pair of coveralls from rough hemp halters. Malcolm squirmed and twisted the coveralls over his legs and torso, a heavy-cotton body sheath that would become uncomfortably warm as the day wore on. Precious slouched against the back of the bench seat, eyeing Malcolm with disingenuous nonchalance as he readied himself behind the steering wheel.

The ignition key elicited an unproductive cough from the starter. Pumping the gas pedal, he said, "What, too early in the morning, Lucille?" The engine groaned into life. Malcolm glanced at Precious through the corner of his eyes as the dash board lights wavered into a green glow.

The gas gauge reached half full. Good enough for now. The aged Scout backed out of the driveway, scudded downhill into the sleepy town of Salty Dog, cruised through the red flashing light and cut left up the valley.

Lucille tracked along the center line of the serpentine pavement, passing beneath tunnels of ancient maples and oaks. Precious sat quietly, her head swaying in rhythm with each shift in the Scout's momentum. Driver and passenger sped by signs and billboards extolling the merits of a devout life. '**Prepare to meet thy God**' was posted on a particularly sharp curve. Malcolm gave a crooked smile and observed. Especially during icy or foggy conditions…Dear God, I should have made that coffee.

A pale rose disc burgeoned through the dust on the rear window, a sure promise of another warm, humid day. The ascending sun lightened the solemn ridges on either side of the valley with a green-grey haze. Malcolm's thoughts wandered from the robotic drive. 'Why did I choose a career that demands so many early morning calls? What drives dairy farmers to start their day so early, motivation, or tradition?'

Malcolm let his foot off the gas pedal and turned into a dirt lane, marked by a powder blue mailbox smeared with glossy black letters. He halted at the top of the lane. From his vantage point, he could only see the house and barn, up on a rise two hundred yards off the road. No rain had fallen in the valley for two weeks, the curled leaves of the foot-tall corn to Malcolm's left suggested a drink was overdue. On his right, cut alfalfa stretched out in long rows for another day of basking in the heat. Malcolm regarded Precious, her front feet on the dash, tail wagging. "Well, girl, I guess we should go see what Happy John has in store for us so early in the morning. Cases here always seem a little out of the ordinary." He punched the accelerator and Lucille tore up the lane, an ochre cloud swirling behind.

Lucille and her entourage of dust settled in front of the barn. A small diesel engine labored behind the milk house, morning milking was in full swing. Retrieving a stainless steel pail through the back hatch, Malcolm scratched Precious' head as he passed by the passenger window. "Stay here, meathead." Malcolm strode into the milk house, a small block building illuminated solely by two windows covered with crisp gingham curtains. Numerous flies struggled on a strip of waxy paper suspended from the ceiling. Malcolm considered the futile efforts of the black specks on the amber-colored mire, wishing there were a few more.

The sink had one handle for the faucet, offering only one type of water, cold and mountain spring-fed. Malcolm chilled his hand while aimlessly stirring the water in the pail, thinking of stirring coffee in a mug. With the pail two-thirds full, he left the damp milk house and re-entered the brightening day. Setting the pail next to Lucille, he massaged the back of his neck with his cooled hand. Precious poked her head out the driver's window. "Still not sure where we're going, so hang loose, sugar pie." Precious planted her chin on the door and sighed as Malcolm vanished into the stable.

A pint-sized terrier yipped and circled around Malcolm's feet as he threaded his way in between two lines of swishing black and white tails. He targeted three adjacent cows, all of which were straddling milk buckets between rear legs. Two lean young men, attired in white shirts and suspenders supporting black pants, and a young woman, wearing a simple blue cotton dress and small white bonnet, washed udders with rags as their bare feet avoided dancing hooves.

"Morning. Where might I find…"

"Oh, it's you veterinary!" A barrel-chested figure squeezed between two cows behind Malcolm, shoving their hips apart with forearms the size of Malcolm's calves. A galvanized feed scoop held in a meaty hand rose in salutation. "Come here early enough to help us milk, I see."

"Sure, John, maybe after I take care of the down cow."

John Y. Peachey beamed, "Samuel, *Herr Tierarzt* wants to help milk our *liebe* Tulip this morning." The older of the two sons smiled broadly at the inside joke. Malcolm surmised that Tulip was a force to be reckoned with at milking time, perhaps light of feet. "*Komm junge Tierarzt*, I'll take you to your patient, we'll need your car," John Y. set his scoop on a straw bale. "*Komm,* young veterinary." While exiting the stable, the farmer called out to Samuel, the German too fluid for Malcolm to understand. John Y. then strode to the Scout. "*Ja, Herr* veterinary, we'll take a little drive up the mountain today."

Malcolm opened Lucille's hatch and secured the pail of water in a box. The suspension rocked with a seismic roll as the farmer bounced onto the passenger seat. To Malcolm, a farmer's weight was proportional to their children's contribution with the daily chores. John Y. must have been getting lots of help.

John Y. examined Precious. "You have a helper this morning, *Tierarzt*."

Malcolm slipped behind the wheel, "I take her out on emergency calls sometimes"—he glanced across the bench seat—"especially when I have to deal with difficult clients."

Precious thumped her bushy tail and licked the farmer's calloused hand. John Y. stroked Precious' pate between her half-erect ears. "Not much of a watch dog, then." Banging his hammer fist on the side of the car below his window, the farmer said. "Forwards, *Tierarzt, Vorwärts!*" Malcolm drove behind the barn, through an open gate and along a worn path that snaked up the pasture.

The engine whined on the steep incline, while Malcolm scanned for imbedded rocks in the lush grass. "She should be just around that next brush pile." John Y. pointed to the right. Malcolm halted at the appointed place. His patient was nowhere to be seen. John Y.'s eyebrows lowered as he rubbed his heavy brown beard between his thumb and forefinger. "Maybe she has other ideas, so she does." With a shove on the door, the farmer pushed himself out to search for his wayward cow.

Malcolm sat with the engine idling, drumming his fingers on the wheel. He hypothesized on how much of his career was spent driving, or trying to find and restrain his patients. About twenty-five yards away, John Y. whistled and pointed to a tree. Malcolm pulled Lucille to within twenty feet and stepped out to appraise the cow sprawled on the grass. John Y. said, "*Ein Rücksack.* Now we'll get some work out of you since you won't help milk the cows."

Malcolm approached his patient. The cow was flat on her side, moaning, on the ground behind her was a glistening red mass. He frowned. Prolapsed uterus to boot, just another day in paradise. "I think I would rather milk the cows, John."

Malcolm returned to Lucille, stumbling on an unforeseen rock lurking in the grass. "Damn it!" he said upon realizing half the water in the pail had splashed out during the bouncy jolt up the hill. Gathering his gear, he instructed Precious. "Better stay there. Can't afford to increase the anxiety of my patient, she already has enough problems." Precious slumped down and laid her head between her front paws. Malcolm called out, "John, is the bull around here?"

"*Nein,* he's in the barn."

Malcolm muttered, "Good riddance." He studied his patient. "I didn't account for the bonus prolapse, John. I'll need you to hike back down and please get one of your sons, a pound or two of sugar, and a good clean feedbag while I treat the milk fever."

"*Ja, Tierarzt.*" John Y. strolled over to the driver's door and grunted his way behind the wheel of Lucille. The engine came to life. "Faster to drive!" Precious sat up and licked the farmer's face as he hollered, "Don't let the cow run away on you!" The Scout pulled into a tight turn. Precious draped her feet over the rim of the passenger window, wagging her tail and looking back at the scene in the pasture. Malcolm watched their progress. There's loyalty at its best, he thought. He clenched his teeth each time he heard his vet mobile scrape bottom. He returned to the cow and kicked the same rock that tripped him before.

The cow lifted her head and attempted to gain her feet but only managed to flop on her side. "Easy girl," Malcolm said. "Let's not tear the uterus." The cow groaned, lamenting her condition and isolation from the herd. Malcolm knew that lonely cows were nervous cows, and nervous cows were unpredictable cows.

Malcolm rested the pail of water on the ground where it was somewhat level. The stainless steel gleamed in the sunlight. He reached out his hand, pushing a halter along the top of the cow's head. His unappreciative companion butted his leg with a quick flip of her neck. "Cut it out," Malcolm scolded, grasping the corner of her mouth with his fingers. He teased the halter over the cow's ears and with a quick jerk, tightened the knot against her jaw. Wrapping the end of the rope around her hock, veterinarian and thirteen hundred pounds of stubborn bovine resolve vied in a tug of war, until her nose almost touched the rear leg. Tying a slipknot, he gently stroked her neck with his hand, "Now just stay cool."

Satisfied with his diagnosis after a quick physical exam, Malcolm yanked a calcium bottle out of his rear pocket, pried the cap open, and shoved an IV hose over the top. Probing in another pocket for a needle, he squatted next to the cow's head. "Relax. We'll get you up in a bit." Blood splattered through the needle and flecked his hand. Clamping the bottle upside down under his chin, he inserted the hose into the crimson fountain.

The flies were yet torpid in the cool of the morning. Soon, the warming sun would incite them into biting cow and human alike. It was a behavior that Malcolm loathed. He was convinced that the level of purgatory set aside for cow doctors included flies, legions of obnoxious, irritating flies. Malcolm stretched as the bottle chugged along, gazing down the slope and surveying John Y's green corn field.

A rock pile larger than Lucille was banished to a corner of the field by the tree line. Every farm had similar shrines, monuments that defined the reality of farming, the struggle to harness the sun's energy while fending off the relentless entropy of nature. The conflict was a delaying action at best, a grind against wind and flood, drought and frost, invasive vegetation, and ravenous pests. When fortune was kind, you had the delusion of running even in the race, but no one ever got ahead.

Malcolm drained a second bottle under the skin and untied the halter. He searched the distant barnyard for the farmer. Since he helped himself to the ride, he should've been back here by now.

Malcolm glanced at his pasture companions, half-a-dozen yearling heifers. Standing abreast and wary of Malcolm's intrusion, keeping their distance. A deep rumble soared from behind the bovine phalanx as an immense black head, streaked with a white blaze of curly hair, towered over their backs. The titan turned its head to gain a better view of the interloper among his harem. The pastoral setting had lost its appeal.

CHAPTER III

Precious sniffed the bag of sugar lying in the middle of the seat while John Y. rested a brawny, tanned arm on the steering wheel. The farmer watched the cows meander from the stable after milking. He reflected, cows go at their own speed, so they do. Samuel walked behind the herd and pierced the barnyard with a shrill whistle, urging the laggards towards the pasture gate. The farmer mused he's almost ready for his own herd. Ellen, a mostly white cow with small black spots, placidly stood while a herd mate mounted her back.

The farmer called out in German, "Samuel, keep Ellen back, she's standing in heat. You and Eli put her in the bull pen then come up to the pasture. We need your help with the calf bed."

John Y. patted Precious' head. "Okay, time to go see your boss." The engine idled while Samuel cut Ellen off from the herd and waved his arms to get the reluctant cow to re-enter the stable. Lazy woolen puffs drifted above the ridge on the far side of the valley. John Y. hoped rain might come this evening, the corn could use some. He frowned, perhaps not so good for the down hay.

As the last of the cows funneled through the gate and dispersed into the pasture, John Y. put Lucille in gear and inched after them. Precious got to her feet, ears cocked. Whimpering softly, she popped her head out the passenger window.

"Haven't you seen cows in a pasture before?" Precious looked back at the driver and barked in staccato beats. "*Was ist los?*" Precious ramped up the frequency. "Well, if you're in such a rush, go back up the mountain on your own legs."

Precious laid her front paws through the window and propelled herself onto the ground. Ears flat against her head, she burst through the gate, exploding the herd in all directions as she rushed through their cumbersome mass and up the slope.

"That dog needs some manners with cows," John Y. muttered.

Eli ran up to the car window, hat missing. The younger boy panted, "Daddy, the bull! Can't breed Ellen!"

"*Was?*"

"He must have got out last night and into the pasture!"

"*Mein Gott!*" John Y. shoved his leg into the pedal. Clots of dirt showered behind Lucille as he grasped the steering wheel with an iron grip. Lucille swerved through the gate, and with a loud snap, the driver's side mirror clipped a post and careened off the door. John Y. roared, "Samuel! Eli! Grab on now!" His sons jumped onto the rear bumper and grasped the luggage rack as John Y. gunned the Scout up the incline.

*****

Malcolm deemed dairy bulls to be jealous, unpredictable, and prone to psychopathic behavior, with the added dimension of a steam roller-sized body fueled by testosterone. Malcolm eased away from his down cow. The bull plowed through the heifers with methodical steps. The heavy chain that hung from the brass ring in his nose dragged along the grass. With lowered head and arched back, the giant called his two legged challenger to a duel with a voice that rang across the entire pasture.

Malcolm's tongue stuck to the top of his suddenly dry mouth. He backpedaled towards a sturdy oak with a forked trunk, twenty feet away. The bull matched Malcolm's steps, sizing up his trivial foe with a deep hum from his chest. Malcolm spoke in a tremulous voice. "Easy big guy, I was just leaving."

With a bellow that echoed in Malcolm's ears, the beast plunged forward with head lowered. Malcolm dropped his empty bottles and ran headlong for the tree; thankful he hadn't put on boots over his sneakers this morning. Scurrying under a limb, he spun to face the bovine tsunami from behind the crotch of the oak.

The entire tree trembled as the bull's poll banged into the trunk. Malcolm crouched behind the barrier, his heart racing. The bull roared and butted the tree again. Stunted six inch horns struck the bark with a percussive knock. Malcolm thought, thank-God his brains don't match his size.

The bull took a step back, his eyes directly even with Malcolm's. The yard-long chain attached to the nose ring shook with each shake of his head.

Malcolm eyed the chain and summed up his odds to wrap it around the tree. He reconsidered. Easy way to lose a hand.

The bull targeted the tree again; the impact landed more to the side of the trunk. "Damn, he's figuring it out, Malcolm muttered. He searched for a rock.

The bull glowered at Malcolm between the twin trunks, frustrated with the standoff. Rolling his head, the chain fanned the air, a show of arms before the next attack. A brown blur flashed along the underbrush and the bull erupted into with rage, causing Malcolm to stumble back from the shock. The beast sprung into the air, flailing the chain and snapping a two inch branch from a limb like a scythe on grass. Simultaneously, he lashed out a rear leg, with Precious locked onto the hock. The dog somersaulted into a thicket of briars. Before the bull had time to regain his balance, Precious sprang onto her feet and latched her teeth onto the rear leg again. The bull thrust his colossal skull through the gap between the trunks, a crazed beast with eyes of malice fixated on Malcolm. Precious forced her adversary to push further into the space. Stomping and snorting in blind fury, the horns became ensnared in the crotch of the tree.

Malcolm snatched the chain and whipped the links about a large limb overhead. The monster bawled, pure spite spewing from his lungs; but with each violent tug on the chain, the nose ring's leverage tightened. Holding on to the chain for dear life, Malcolm cried out, "Precious! Enough! Over here!" Precious released her hold and slunk over to Malcolm, wagging her tail.

Malcolm bent over, his hand shaking. "Thay will do!!" His fingers found a small amount of blood along her back. He clasped her neck as Lucille came to an abrupt stop ten feet away.

John Y. rolled out of the car, his sons holding onto the rack with white knuckles and hair tousled. "Are you okay, *Tierarzt?*"

Malcolm's unsteady voice barely carried over the bovine bluster. "I thought you said the bull was in the barn."

John Y. examined the combination of the bull's head, chain, and tree. "Maybe got through the fence without minding the fencer, so he did."

Malcolm nearly retorted, "The beauty of artificial insemination, no damn bulls on the farm," but kept his opinion within. "John, can you come over here and hold this chain? I'm going to get a tranquilizer to knock him onto the ground."

John Y. cautiously approached Malcolm and gripped the chain. "You sure got him, veterinary."

Malcolm reached down and stroked Precious. "She deserves the honor, she blindsided from the rear and got him to pin himself between these tree trunks."

John Y. rubbed his beard and surveyed Precious. "Your friend has spunk for such a little thing to take on something his size."

Malcolm looked into Precious' eyes. "Yeah, she's special."

"What kind of a collie dog is she?"

"Don't know for sure. Got her from an animal shelter; some sort of Aussie shepherd crossbreed."

Malcolm slapped a dose of tranquilizer into the bull's neck muscle, ignoring the violent protest. Several minutes later, the bull relaxed while John Y. forced the head back through the tree. The bull swayed and stumbled a few paces, slumped to his front knees, then lowered his back legs. His deep voice softened into snoring. Malcolm welcomed the peaceful sound. "Now we can deal with the prolapse."

John Y. plucked the straw brimmed hat off his head, wiped his wet brow and surveyed the cow. She remained where Malcolm had left her, resting on the grass by the pail of water. The farmer swept a hand as if introducing his sons onto a stage act. "I thought my boys could help."

Malcolm regarded the pair with a thin smile. Farmers perpetually alluded to their sons as 'boys', no matter how old or experienced. "John, would you please slip that feedbag under the calf bed? A strong whiff of sweat descended on Malcolm as the farmer bent over and hoisted the uterus onto the feedbag, suspended like a hammock. The large chest cavity of the cow resounded with the slap of Malcolm's hand. "Come on girl, get up!" The cow rolled onto her front knees and elevated her wobbly rear legs, while Malcolm lifted her by the tail. He admired the will of the creature to gain her feet. Not bad considering she had given birth, collapsed into hypocalcemic shock, and had her uterus dangling out her hind end.

John Y.'s sons grappled the cow to the nearest tree, wrapping the halter about the trunk. The farmer watched Malcolm administer an epidural while he leaned against the rump of the cow.

"Keep her from pushing against me," Malcolm explained. Holding the pail in one hand, he splashed off bits of grass and dirt from the uterus with handfuls of water. He shook the bag of sugar over the glistening tissue, causing red fluid to seep from the surface and the uterus to shrink in size. "All set, John."

John Y. cheerfully volunteered his 'boys' to suspend the uterus with the feedbag, while he assumed halter duty. The bull snoozed under the tree while Precious lay in the shade of a shrub, licking her legs.

Malcolm pulled a plastic sleeve over each arm and slowly coaxed the uterus back into place. He got into a rhythm, three inches in, two inches back, hold, and push again. Sweat stung his eyes while flies buzzed about his face and bit his ears. Arms and hands encumbered, the best he could do was jerk his head or blow air at his tormentors in futile puffs.

The herd had encircled the cow in a show of curiosity and support, their eyes and ears focused on Malcolm. A pair of scouts sifted closer and stretched their necks just out of arm's length to gather the scent of the metal pail. Precious sprang from her respite, growled, and bared her teeth. The instigators retreated to a respectful distance.

John Y. chuckled. "She's a good watchdog after all, *Tierarzt.*"

Malcolm winced as he pushed against the uterus. "She keeps an eye on my back."

Halter in hand, John Y. offered numerous quips and banter from beneath the shade of the tree while Malcolm answered in monosyllabic grunts. For the second time this morning, Malcolm questioned his career path. Finally reaching the critical balance point, he pushed the uterus over the hump in the pelvic floor and distended it into the abdomen. With a triumphant breath, he yanked the sleeves off sweat-soaked arms and straightened his stiff back. A fragment of his coverall collar that was still dry served to sponge the stinging salt from his eyes. "Now she feels better," John Y. commented. Don't we all, Malcolm contemplated as he watched John Y.'s sons, liberated from lifting detail, wipe trickles of sweat from their smooth-shaven faces.

Malcolm looked about him and exclaimed, "Get out of there!" Several cows had gathered around his Scout, one of them with her head through the driver's window. Strands of saliva glistened on the dashboard and steering wheel. Precious took her cue and leapt to Lucille's defense, harassing the vandals into a canter. Malcolm shook his head. "Cows are nothing more than a bunch of oversized children."

The farmer laughed. "I guess your *Hund* can't watch everything." Turning to his sons, the farmer said, "*Samuel und Eli, bringt die cow into the stable, bitte.*" With that, the young men untied the halter from the tree and slowly led the unsteady cow down the slope. Malcolm gathered up his things and examined Lucille. "What happened to the side mirror?"

"No problem, *Herr Tierarzt,* it came off on one of the gate posts in the barnyard. You get it fixed and I'll pay the bill, might be the only new thing on that car of yours."

Wearily placing his gear in back and shutting the hatch,

Malcolm slumped into his seat, with Precious once again squeezed between him and the farmer. Lucille crept down the slope behind the haltered cow, who was disinclined to move with any faster than a casual walk. John Y. pushed himself out the door when they re-entered the barn yard. "I'll close the gate and see you at the milk house, veterinary."

The milk house was an oasis of chill comfort amid the growing heat of day. Delving his cupped hands under the faucet, Malcolm splashed copious amounts of water onto his face and scalp. The water electrified his skin, dripping down his neck and under his shirt. Refreshed, he returned to Lucille, emptied his pockets and sorted out the debris into a trash can behind the passenger seat.

Precious stretched her neck and sniffed his face, intrigued by the scent of uterine fluids. Malcolm groused to himself. Nice start to the day, no coffee or shower. Beating her tail on the seat, Precious focused her attention through the passenger window. Malcolm turned and beheld a plate of oversized cookies.

"It's a good thing we didn't use all the sugar for the cow," John Y. said while helping himself. Malcolm hand-picked one of the yellow discs and savored the rich mix of sugar and butter in his mouth. If this was a regular dietary habit, he understood where John Y. acquired part of his girth and boundless energy. The 'boys' appeared from the barn, returned the halters and snagged a couple cookies for themselves.

John Y. beheld Precious. "*Was ist der Hundername?*"

"Precious."

The farmer offered a cookie through the window. She gently grasped it with her teeth. He nodded, scratching one of her brindle-colored ears. "A cookie for the bull catcher."

"Is all your hay down, John?"

"No, we took in half our field last week."

"What about the rest?"

John Y. mulled over the parallel green rows stretching across the field. Calculating eyes lifted to the sky. "Probably won't be dry enough until tomorrow, unless it rains tonight, then we rake and dry again."

"That won't do much for the quality."

John Y. took another cookie; half gone in one bite. He shrugged. "Can't do much about the weather." He continued to scrutinize the hay field. "Hmmmm, the dark ones have found something to their liking." Several crows were pecking at the ground, posturing and hopping about their discovered treasure.

"What are they after?"

"Dead mice caught in the hay mowing."

A companion to the black scavengers lighted onto a limb hovering over the farm lane, calling out noisily. "A sentry for the feast," Malcolm surmised.

"And that makes five, not so good."

"How so?"

"Sickness comes with five."

"Maybe a sixth one will come by to even it out."

John Y. frowned. "Six is a funeral."

Malcolm handed John Y. a bottle of penicillin and an invoice. "Thirty-five cc's twice a day for three days for the prolapse patient, John."

Holding out the plate of cookies, the farmer said, "You better take another one in case you get lost on the way to the office." The 'boys' grinned.

Malcolm's stomach rumbled as he grabbed a cookie, reminding him that he hadn't eaten breakfast. "If she goes down again, give us a call."

Driving up the lane, Malcolm beheld the broken mirror that John Y. had laid in the passenger seat. Floyd and Lloyd would have to perform their magic once more. Purple and lavender phlox swished along the roadside in the wake of Lucille's headlong rush to Mayville. Malcolm regarded Precious. "Miss Lois will want a damage report about you."

CHAPTER IV

Life had already shifted into full gear by the time Malcolm entered Mayville. Buggies, pickups and farm equipment churned about the streets while burnished stainless steel milk trucks lined up at the cheese plant to deliver the morning's catch. Malcolm turned onto a maple-lined side street, piloting Lucille to a flat-roofed, one-story brick building. He parked between a burgundy Jeep Cherokee and a blue Ford 4x4 pickup. Malcolm stretched onto the asphalt and jerked his thumb towards the clinic door. "Come on, we need to take a look at you." Precious scampered out of Lucille, leaving the Scout looking tired and faded compared to the newer vehicles on either side.

The aluminum screen door groaned a tired welcome as Malcolm stepped onto a green linoleum floor, scented with Tincture of Green soap. The walls were obscured by pine shelves filled with bottles and vials, except for two grottos on opposite walls that allowed enough space for a wooden bench, the entirety of seating for waiting clients. By virtue of a desk, telephone and adding machine, the room also served as the office.

As Malcolm continued into the backroom, a baritone voice hailed his arrival. "Good morning, Malcolm."

"Morning, Henry."

Holding a pot of freshly brewed coffee, Henry Breck raised an eyebrow. "Your canine assistant was on call with you this morning?"

Malcolm rummaged for a carton of milk in the refrigerator among vaccines, microbiological media plates, and tissue specimens awaiting laboratory submission. "A milk fever with the added bonus of a prolapse at John Y. Peachey's. Precious, and a sturdy oak tree, saved my neck from a bull in the pasture." Malcolm splashed some coffee into a mug. "For that matter, she probably saved every bone in my body."

Henry hesitated from reaching for mugs hanging under the cupboard. "Bull?"

"I guess the brute blundered through John Y.'s electric fence last night." Malcolm petted Precious. "She got a little roughed-up, I wanted to check her out."

"Mmmm." Henry studied Precious. Henry reminded Malcolm of a medieval baron, stocky, clean-shaven head, deep brown eyes, and a close-cropped black beard and moustache with a hint of grey. The two-inch scar on the top of his scalp from a mishap with a cattle chute fostered Malcolm's illusion, perhaps a jousting injury in a former life.

Henry peered into a brown paper bag, selected a chocolate-iced ring, wrapped it carefully in a paper towel and placed the bundle into a microwave. Fifteen seconds later, Henry recovered the sizzling doughnut and hot grease teased Malcolm's nostrils.

Malcolm searched the bag and extracted a jelly-filled Bismarck. He pulled the door to the microwave. "You've gotten me addicted to doughnut nuking."

Henry shrugged. "They're better that way." Henry meticulously poured coffee and balancing his treat, two mugs, and the doughnut bag in his hands, walked into the front room.

A hand thrust through the back door that led to the storeroom and plucked the ringing phone. "Good morning, Mayville Veterinary Clinic."

Malcolm sat down on the bench opposite from Henry, washing down the near-scalding jelly oozing in his mouth with a long overdue slug of coffee. Precious sat directly in front of him, eyes fixed on the prize.

Henry inquired, "So it was black topper John Y. Peachey over by the front mountain?"

"No, Happy John Y. of the yellow buggy clan on Rocky Creek

Road.”

“His usual self was he?”

“Let the good times roll, all day, every day.” Malcolm slipped a piece of doughnut to Precious, whispering, “Don’t be such a mooch.”

A brunette with short hair and an agreeably mature figure tucked underneath a pullover shirt and tight jeans appeared from behind the storage room door with the phone to her ear. Her hand oscillated on the desk surface as she scrawled notes on a piece of paper. Without looking up as she replaced the phone, “Morning, Dr. Malcolm. Are we ready for another day?” Precious sauntered over by the desk and rolled onto her back.

“I think so, and you Lois?”

Precious spread her legs as Lois briskly rubbed her belly. “How’s my favorite girl?” The impromptu canine massage halted and Lois’ eyes narrowed. “Do you ever brush this dog?”

Lois Knepp was an aggregate of office personnel; book keeper, accountant, microbiologist, pharmacist, drug procurer, mail clerk, animal handling specialist, custodian, and as such, indispensable for the practice.

“She’s all matted from a tussle with a bull in John Y.’s pasture this morning.”

“Bull? A bull attacked her?”

“He was actually after me.”

“She could have been hurt, poor thing.” She pulled the lid off a heavy glass jar on the desktop and offered a dog biscuit to Precious. “She needs to be cleaned up and bathed.”

“I was going to look at her.”

Lois grabbed another biscuit. “I’ll take care of this while you’re on the road today, more likely to get done right.”

Sipping his coffee, Henry interjected, “Anything else carrying over from the weekend, Malcolm?”

“Not really, I told John Y. Peachey to call if his cow goes down again. He’s pretty easy going about things like that.”

Henry said, “Comes by that honestly, much like his father.”

Lois added, "Old Sam Y. Peachey, does he still live and work with John Y.'s brother Noah Y. at the harness shop?"

"Yes, retirement Amish style."

The phone rang again; Lois seized the handset as if it was falling off the desk. Precious lay on her back, eyes closed, legs curled in the air. "Which reminds me Henry, I've got to check with Floyd and Lloyd this morning."

"What's the problem this time?"

"I think the starter may be going…and John Y. had a mishap while driving into the pasture."

"Driving in your Scout?"

"Yeah, he clipped off the side mirror at the gate."

"Good thing he isn't likely to apply for a driver's license anytime soon."

Lois finished the call and fidgeted with her pen. "We have quite the list today." She ran a length of paper from the adding machine and tore it into two pieces. Henry and Malcolm approached the desk, receiving their assignments. Malcolm decoded Lois' scrawl, ensuring he knew each of the farmers, and more importantly, the location of the farm. Nearly two years in the valley and his learning curve on local farm geography had yet to level off.

Lois' face looked as if she bit into a tart lemon, "No shower this morning?" Precious opened one eye and peeked at Malcolm.

"The cow did have a prolapse, Lois."

"Shower yesterday?"

"What? Should I shower more than once a month?"

Lois sat back in her chair, arms folded. "Did you guys get your invoices filled out over the weekend?" Henry and Malcolm glanced at each other. "You know the end of the month is coming up, I need those ASAP! Henry, ever since you bought the practice from Paul, the paperwork has been way too lax around here. And since you were the one that hired Malcolm, you're responsible for his bad habits."

Henry cleared his throat, "Now Lois, we'll get caught up."

"We still have to figure out the accounts receivable. Why the other day…"

Malcolm escaped into the back room and retrieved another scrap of paper out of his pocket, his 'shopping list' to restock Lucille. He washed his mug in the sink and pulled an assortment of bottles and supplies from the shelves and filled his cooler with vaccines and an ice pack from the refrigerator. Re-entering the front room, he found a new arrival sporting a trim beard and straw hat with a narrow brim that was upturned in the rear. Raymond L. Yoder was chatting about the weather with Henry and Lois. Malcolm offered the doughnut collection to the farmer.

Raymond L. smiled on seeing the bag. "Good morning young veterinary." He tunneled his large hand into the bag and selected a glazed ring.

Inspired, Malcolm hoisted another Bismarck for himself. "Good morning, Raymond. How are things?"

"Why, I was just telling Lois and Henry that we could use a good rain, last week's didn't give us more than a half inch."

"I was at your brother Kore's last week, and he said he got nearly a whole inch."

"The rain can be that way, so it is. He even got a shower that my farm didn't three weeks ago."

"Huh, they must lead a more righteous life over on that side of the valley." Raymond L. knitted his brow as Malcolm made for the exit.

Lois was perched by the door, jaundiced eyes cast on his second doughnut. "How many of those have you eaten this morning?" Snatching the contentious lard bomb, she stated, "Get some fruit from a farm stand."

Momentarily stunned, Malcolm recovered from the ambush, "Hey, where you going with that?"

Lois handed the plunder to Henry. "It won't matter to him, he's over-the-hill." She turned and smiled disingenuously. "Besides, we have to keep you in shape to attract the local farm girls." Still upside down, Precious wagged her tail slowly.

"I can do without that thought, Lois. Much as I would like to chat about my future relationships, I don't have time to talk about that now. Besides, I need to get more gas for the road this morning."

"Maybe your new father-in-law will be a Peachey, Byler"—Lois shot a glance at Raymond L.—"or a Yoder." Malcolm retreated through the door. She called out, "Provided that you bathe. I can hose you down later today after Precious."

Storming past Raymond L.'s black pickup, Malcolm muttered, "What a handful, no wonder she got divorced." He slammed Lucille's door and stomped on the accelerator, reaching the gas station in record time.

Lucille's tires rang the pneumatic bell, and a teenager appeared by Malcolm's window like a genie summoned from a lamp. "Fill it up with regular please." Malcolm grunted as he tugged on the reluctant hood release handle. Leaving the kid cleaning the windshield, Malcolm walked past the Coke machine and into the garage. The rear end of dark blue cotton work pants anchored by brown Redwing shoes protruded from under a hood of a Pontiac station wagon. "Morning Floyd, wonder if you could help me with the Scout?"

The figure twisted around. Periwinkle-tinged eyes squinted through thick glasses, leering from a shadowy face beneath the trouble light. "What was that, Doc?"

"I need a new side mirror for my Scout."

Floyd's voice drifted, "Mirror?"

"My side mirror came off the Scout during a bump into a fence post."

"What in the Sam Hill were you thinking? You know parts are hard to come by for Scouts. Jeez oh Pete, Doc! You should take better care of that vehicle."

Malcolm opened his mouth to respond but changed his mind. He retrieved the mirror from behind his back and offered the part to Floyd.

"Holy Mother of God, Doc! You smashed this to bits. You veterinarians race around the roads like you're in a chariot race."

He yelled, "Lloyd! Where can we get a mirror for Doc's Scout?"

A voice carried across the grimy floor from underneath a small dump truck in one of the other bays. "Did you say mirror?"

"Yeah, side mirror. Doc's gone and knocked his side mirror clean off that Scout with the goofy patchwork body."

"What year was that?"

Floyd faced Malcolm. Malcolm mumbled, a penitent sinner, "'76."

"'76," Floyd echoed.

There was a pause, "Might have to call Leitzinger's."

Still sputtering under his breath, Floyd grabbed a rag and wiped his hands. "Follow me, Doc." Floyd paced away, putting his hand on a hoist cable. "Don't whack that tall head of yours. Of course, it might knock some sense into you." Malcolm bent over and followed the mechanic into a cluttered room that smelled of old grease.

Floyd flopped into a leather swivel chair that had seen better days and creaked across a threadbare, stained carpet remnant, the original color no longer recognizable. Displacing a carburetor and sheaves of loose paper, Floyd found a yellow writing pad on the desktop. He smudged an oily thumbprint into the paper as he flipped the top page over. "Alright, let me write this down. What year?"

"'76."

"It'll probably cost you damn near twenty bucks for the part, don't you know."

Malcolm asked, "I suppose I'd need the mirror for this year's inspection?"

Pedantic eyes laced into Malcolm over the top rim of the glasses. "Now what do you think, Doc? The citizens of this state deserve to have people like you drive safe cars on the road, so they do, especially the way you drive." Floyd scribbled on his pad. "Damn International going belly-up and all, kind of dropped the ball on supplying Scout parts don't you know."

"Yeah, dropped the ball," Malcolm replied, admiring a reclining blonde on a creased Playboy calendar on the wall.

"By the way, Doc, aren't we due for that inspection soon?"

Malcolm nodded. "Which reminds me; I'm still having issues with the ignition. Getting more unreliable, I think the starter may be going."

"Or the ignition switch, or the distributor cap, or a bad connection to your battery. We'll take care of it."

Malcolm found the kid in the office, paid for the gas and escaped to Lucille.

*****

A long, hot day done, Malcolm started the grill and released himself from the day's labors with the overdue shower. Wearing only a pair of jeans, he searched for a John Lee Hooker LP from a milk crate full of vinyl discs next to his aging turntable. He flamed a couple of hamburgers, sipped a Yuengling and ate on the patio. "You ain't nothing but a hound dog," he hummed, tossing a tidbit of ground round that disappeared in midair between tongue and teeth. "I'll bet Lois spoiled you rotten today."

Scores of night stalkers harmonized a chorus of trills and screeches, replacing the blues as the evening wore on. Precious cocked her head and stood by the door. "Woof!"

"What's your problem?" Malcolm scanned the western horizon, ominous clouds masked the receding twilight. Was that a short glimmer of light in the distance? Lightning perhaps. Bursts of light erupted from the clouds more frequently. "I hope the storm blows this damn heat into the next county," he thought out loud. .

Precious crowded past him as he opened the screen door to get another beer. She remained indoors as he returned to the patio. Malcolm smiled at her absence. Must be the storm is heading this way. The ridges volleyed rolls of thunder across the valley to one another, playing sonic ping-pong. Despite the impending weather, the languid air kept him in his chair.

"Woof!"

Malcolm spoke through the screen window. "Yeah, I know you don't like thunder, wingnut." He pushed his stiff back out of the cushion and covered the grill. Today had been another arduous day, like so many others that consumed his life with a voracious appetite. "Let's call it a day." Precious bolted straight to the bedroom.

Malcolm undressed and retrieved an old Raymond Chandler paperback on his nightstand. Precious jumped onto the mattress and settled by his feet. Flipping through only four or five pages, he pulled the light switch and lay with his hands cupped under his head. The storm pushed forward, heralded by trembling leaves in the hickory that towered above his back yard. Riding on a gust of wind, a shrill bolt fractured into feral howls and ricocheted off into the night. Tentative drops of rain tapped on the gutter above the bedroom window and hesitated. Another rush of air animated the curtains. Precious scurried to his side, trying to bury her head under his pillow. Malcolm stroked her back. "It's alright."

CHAPTER V

A drab cloak clung to the shoulders of the ridges, grey remnants of the previous night's deluge. Wisps of fog levitated, wavered then descended along the spurs of the mountains. Fat drops of rain splattered Lucille's windshield as Malcolm wheeled into Dale Calhoun's barn yard. The rain pelted his back while he retrieved his equipment through the back hatch. Hurrying into the stable; he came across a heavy-set woman with strong arms and shoulders sculpted from a dual career of farm work and motherhood.

"Got a little wet did you, Doc?"

"Little bit, Sally."

"Little bit? More like a drowned rat," added Dale, Sally's husband. Steel-colored eyes, with a tinge of restlessness, stared from under his worn cap. "Where we starting, hon?"

"Over on Elsie."

Malcolm approached the backside of the cow, Dale leaning on her flank. Sally called out, "Fifty days, Doc."

Malcolm reached inside Elsie's rear end in search of a uterus. "Good rain for the corn, I imagine?"

"Yeah, but we got ten acres of hay down that's pissed on. Ain't that good, anyhow, got cut late because we were planting corn late with all them spring rains." Dale tended to look up and away while speaking, perhaps beseeching heaven for clarity and understanding. His left hand, minus the ring finger, courtesy of an entanglement between a cow's neck chain and his wedding band, stroked Elsie's back.

Malcolm announced, "She's pregnant, Sally."

Sally pointed. "Forty-two days, Doc."

Malcolm shuffled to the next bovine derriere in line. Dale continued, "Of course we might have got more hay baled if the damn gear box wouldn't have given out in the baler. I told the old man to buy a John Deere instead of New Holland."

Sally reproached, "Dale, hay equipment of any kind can have problems."

"Not like this, it don't."

"Jim Boyle has a New Holland, he gets nice hay."

Malcolm intervened, "She's pregnant too, Sally." He forged ahead, maneuvering between cows and spousal debate over the pros and cons of farm equipment brands. He mused, I should charge these two for counseling. Finishing with the last cow, Sally suggested, "Let's go in for a coffee."

Malcolm adjusted his eyes to warm sunlight gleaming on the wet grass outside the barn. Holding his hand over his brow, he spotted a wiry figure on the far side of the barn yard, tugging on a socket wrench in the engine of a John Deere 4240.

Dale tracked Malcolm's line of sight. "That's my brother, Jack."

"Don't know if I've met him before."

"Been away for a while. He's staying on our other place up the road. He's handy to have around; crackerjack on repairs for anything that runs on wheels."

Sally added snidely, "I guess green tractors need fixing, too."

Dale curried his bushy moustache with thick fingers and redirected his gaze up the farm lane. "There he goes again, that son of a bitch, third time today."

A white Chevy pickup rolled along the road a couple of hundred yards away. Malcolm asked, "Who is that, Dale?"

"None other than John MacDougall, biggest trouble maker down this end of the valley. Good for nothing, just plain ignorant."

"Oh, now hon, he's just a little to himself," Sally said.

"Does Mr. MacDougall farm, Dale?"

Dale looked at Malcolm as if he was stepping out of a newly grounded UFO in the farmyard. "That pot licker don't but scratch a living raising a steer or two. Ain't worth the tits on a boar. Spends his time drivin' up and down the road all day, seein' what the neighbors are doing and spots deer with his truck at night. Folks say he grows the whacky weed up on his place."

"Now Hon," Sally implored.

Dale was undeterred. "It's true. His brother came back from Vietnam high on dope. He started the whole business."

Malcolm reflected on Dale's neighborly love while walking up the slight rise to the farmhouse. A pristine red-orange '70 Chevelle 396 SS, black racing stripes across the hood, adorned the yard next to the porch; Dale's incentive from his father to work on the farm rather than finish high school. Malcolm remorsefully considered Lucille's patchwork body of putty, primer, and bi-color paint.

The floor of the front porch moaned with the burden of each step on the tired boards. The rambling house was divided into two living quarters, one for Dale's parents, one for Dale, Sally and their kids. Dale barged into the kitchen. "Come on in, Doc."

Piles of mail and advertising flyers covered the woodstove. Canning jars of pickles and jam, bread, a sleeve of saltines, a butter dish, and three or four newspapers lay on the table. Neglected toys were strewn on the floor, alongside a couple of hats—tokens from seed corn salesmen.

Dale stuck his head into the family room. "Hello, boys!" His two elementary school-age sons sat mesmerized in front of the television with a blond Amish boy, captivated by Bugs Bunny. Malcolm's eyes widened. Dale winked. "That's Sammy Yoder, comes here on errands and the like, stays over to watch a little TV."

"His folks know this?"

"Oh, they know. Besides, it ain't *his* TV. You know how them dutchmen use our barn phones, as long as it's not *theirs*."

And old Scouts, thought Malcolm as he slouched into an oak chair at the kitchen table.

Sally pulled out a pitcher of milk from the refrigerator.

Grabbing a ginger cookie and pouring milk into his mug, Malcolm said, "This MacDougall in the white truck sounds intriguing."

Husband and wife looked at one another. Dale responded, "The MacDougalls ain't right, Doc. Some of 'em been mad as a hatter."

"Like the Addams Family on TV?"

"This isn't funny crazy, Doc." Dale leaned closer and actually managed to focus his eyes on Malcolm's. "They're evil," he nearly whispered.

"How evil?"

"Little better than a gang of thieves."

"Going on for generations," Sally said.

"Once in a while, one of 'em just loses their grip on life."

*****

Precious greeted Malcolm at the front door, scolding him for his late arrival with a tirade of barking. "Don't give me so much grief. You know you can't stay in Lucille all day during hot weather. I'm sure Miss Violet saw to your needs." She ran out the door and patrolled the perimeter of the backyard. A rude squirrel taunted her from a limb of the hickory tree.

Balmy air whispered through the bedroom window as the warmth of two fingers of Oban diffused through Malcolm as he lay in bed and recounted his day. Malcolm closed his eyes and pondered yet another new acquaintance in the valley, John MacDougall.

CHAPTER VI

In the valley, Tuesdays were auction days. Farmers packed into the eatery at the Mayville Auction Barn, overlooking the auction ring, squeezed onto folding chairs of brown steel, elbows resting on worn formica tables. Soil-caked work shoes and boots rested on a yellowed and cracked linoleum floor and the entire space smelled of beef and noodles, pancakes and eggs, ham hock and bean soup, and pie. Waitresses with head caps, knee length cotton dresses, and navy blue tennis shoes, sorted through diners conversing about the cost of farmland, cattle, hay, and above all, the price of milk, which was the source of perpetual commiseration.

Tents, trailers, and tarps sprawled over the grounds, beckoning hordes of bargain hunters and hucksters from miles around. Townspeople looked out their windows with disgust at the infestation. Traffic clogged the valley highway behind unhurried buggies, inciting impatient drivers to pass in spastic bursts on blind curves, pressing horns, or shouting at the horses.

To Malcolm, auction days were a pain in the ass. The congested traffic reminded him of the chaos of his hometown in suburban Chicago. Fortunately, Lois had decided to send him over the front mountain this morning to dwell in the river valley to the south. Fleeing the fracas, Lucille strained all eight cylinders through the shaded steep road on the side of the south ridge. The canopy widened into an aperture of blue sky as Malcolm yanked the steering wheel into a small gravel lot and cut the engine. He stepped into the sun, a two-legged speck on the shoulders of a leviathan of rock, gazing from the summit at the world below.

A silver coil shimmered in the morning light, coursing at the foot of the next ridge to the south, the Juniata River. Beyond, verdant mountains undulated towards the horizon. Malcolm turned, strolled across the pavement and surveyed the valley, pondering the patchwork of green and yellow—farmers, stewards of the garden for countless generations; chasing the sun from one end of their fields to another as the earth raced a hundred miles in orbit in three blinks of an eye. A sister ridge bordered the valley to the north, the back mountain. Front mountain. Back mountain. Up and down the valley. Even the sense of direction was peculiar to this place.

Content with his moment of peace, Malcolm sauntered back to Lucille and continued down the grade on the opposite slope from the valley. A large raccoon waddled onto the road in front of Lucille. "Son of a bitch!" Lucille swayed as Malcolm's foot jammed the brake pedal. The sudden pull on the wheel reawakened his throbbing shoulder, bruised from being rammed into a fence by a surly heifer a week ago. Malcolm stole a quick look into the rearview mirror. Dumb raccoon needs his head examined. Unfazed, the creature kept strolling up the road.

The road curved away from the base of the ridge and the Scout lurched into the river valley, the land of pariahs. No buggy Amish farmed here; the mountain was too imposing for visiting kin on the other side. Few 'black car' Amish or Mennonites lived here either. Those that chose to do so evaded the subtle surveillance by their brethren in the righteous valley back over the ridge.

Malcolm worked through his calls, eventually winding along a road that hugged the south bank of the river. The sun was angling towards a notch in the ridges to the west as he climbed over a steep hill, revealing the county seat, Gillstown. Plainly dressed people slumped in lawn chairs on front porches, passing time as the world rolled by their timeworn row houses on the highway. Like a rapacious tramp, the owner of the textile mill by the river had flitted off ten years ago to court cheaper labor to the south, then whored out to even more attractive suitors overseas. The steel mill in town was still operating, but with a third less workers. Gillstown had known better days.

On the east side of town was the Galloway County Farm, a holdover from a time when milk, beef, eggs, and produce were provided for the county homes for the elderly and mentally challenged. Presently, the farm supplied ample work for those who were paying restitution to society for drunk driving and shoplifting. The farm was Malcolm's last stop of the day, his preference. An impromptu happy hour with the herd manager Bobby Casper usually ensued after meeting the needs of the bovine patients.

Bobby was studying his second empty bottle of Rolling Rock, the green glass reflecting light from two large windows overlooking the barn. "So what are you doing next Friday night?"

Malcolm slouched on a worn couch, concealed by an orange afghan to keep the foam from dissipating about the room. "Not much."

Bobby sat back in his armchair, rubbing his ebony beard. "No love interest?"

Malcolm regarded Bobby with a stony face.

"Hey, you gotta get back into it and act social sometime."

"I'm not a social person."

"Our dart team is playing out your way that night, tossing at the Dead Bear."

"As in the bar in Salty Dog? Down the hill from my place?"

Bobby tilted his head. "That's the one."

"You, play on dart teams?"

"Yeah, it's a good drinking game."

"Throwing sharp objects is a good drinking game? How many teams play?"

"Two on a given night, seven or eight in the league. No real schedule, the thing of it is, it's hard to get drunks to stick to anything regular."

"Other than going to the bar, of course."

"Yes, *that*'s mighty regular."

"What time does the fun begin?"

"More or less seven. Maybe you'll come across that special someone."

"I don't think that would be a likely venue for such a meeting."

"Fate is unpredictable." Bobby glanced at a small clock on a dusty bookshelf opposite his chair, "I better go see how the milking is going, got a couple of new associates today."

"Visiting scholars from the gifted educational program?"

Bobby shook his head. "Never thought my animal science degree in college would train me to be a parole officer."

"So why not run your own herd?"

"No money to start my own, besides I like managing this one."

"Even with having to work with all the hand-picked trainees?"

"Today's new arrivals broke into a place, helped themselves to stereo and video equipment, went down the street to their car, opened the trunk to hide the stuff, and locked their keys in the trunk when they closed it."

"Professionals."

"Needless to say, the police arrived during their time of need."

"And you think these boneheads can find the correct end of a cow to attach a milking unit?"

Bobby stood up, ready for barn work in a tee shirt, tattered jeans, and dirty sneakers. "I hope not, they could use a good kick in the shins. Charlie's got an eye on them. He'll make sure they're delegated to milk the cows with light feet."

"You're twisted."

"Doing my part to make Galloway County a nice place to live, work and raise a family."

Malcolm rocked to gain momentum up from the cavernous cushions of the couch. "I'll reserve a box seat for Friday, haven't attended the Dead Bear for a spell."

*****

Malcolm entered the bedroom in the back end of the house, thankful to find no message on the recorder. The room was furnished with dissimilar relics collected from local garage sales, a poster bed, watermarked nightstand and a dresser. Tossing his keys onto the dresser near a picture frame, he ran his fingers over a black and white photo of an attractive blond wearing a simple black dress, an erudite and confident woman in her element. He couldn't remember where this moment had been captured; probably at one of the many well-heeled functions they had attended when they were first married, a world away from where he stood now.

Malcolm picked up the frame, gently cleaned the dust from the edges, and reminisced on his first year in practice. How many times had it been half-asked, half-stated, "You're not from around here are you?"

Malcolm recalled a cold, predawn trek during his first winter. Grappling with Lucille's steering wheel, slipping and sliding on a blanket of virgin snow, he finally reached his destination. The headlights illuminated a pleasant surprise. The farm lane was plowed, an unexpected luxury after the perilous drive. Malcolm gratefully thanked the farmer for the effort on his behalf. The farmer gave the novice veterinarian an enigmatic look and bluntly stated the lane was cleared for the benefit of the milk truck, the carrier of the life blood of the farm to market. After ensuring that Lucille wasn't sitting where the truck needed to park, the farmer pointedly directed Malcolm to the cow that needed attention.

But during that first year, Malcolm endured a more ponderous challenge than stoic dairy farmers, his wife's distaste for life in Galloway County. Her decisive departure perturbed the complex web of conformity in the valley. Two hundred farms, many of them without telephones, but word got around.

Placing the frame on the dresser, he went to find something to eat. Precious, sensing his purpose, walked beside him. Stiffened pieces of pizza with pepperoni and onion lay inside a box in the refrigerator. Malcolm commented, "Well, Precious, its better the next day, you know." She wagged her tail.

They shared supper on the patio, Malcolm resting in a metal chair with chipped green paint. Precious sat at his side, her eyes shifting with each elevation of hand to mouth. Malcolm whispered as a piece of crust vanished from his outstretched hand. "You good for nothing beggar, you finish your own food already?" He scratched her head as she chewed the prize with her mouth open. Irises and peonies reposed in the garden. The north ridge of the valley slanted away to the horizon, a perfect study of perspective and distance. Puffy clouds wandered into the landscape over the front mountain to the south. He smiled; pizza on the patio was far better than the elegant brunches he attended in his former life. The ambience was far less pretentious, and Precious was better company.

CHAPTER VII

The next day, a brisk breeze scoured the clouds off the ridges and the sun took control of the sky. On the way home, Malcolm stopped by Jonas P. Yoder's, marked by a hand painted sign at the end of the lane—brown eggs, dressed chickens, and strawberries, no Sunday sales. He waved at Jonas P. bouncing in the seat of his hay rake behind a pair of massive Belgians. Malcolm regarded the farmer's progress. His hay got rained on like everyone else's, I guess.

Two border collies escorted the farmer, nose to the ground, seeking mice and ground hogs. Barn swallows spiraled about the steady moving horses like dervishes; agents of mayhem as insects hopped, jumped, and flew into the air to flee the ponderous hooves and wheels, only to become easy meals.

A spectacled teenage girl with a black kerchief wrapped about her head entered the barnyard from an immaculate weed-free garden.

Malcolm said, "A quart of strawberries and a dozen eggs please."

She smiled as her eyes briefly met his and padded on bare feet into the milk house. Leaning against the hood of the Scout, Malcolm surveyed the farmstead. The wood siding on the barn, house, and even the outhouse were in need of paint. The two story house had no curtains or screens over the windows. A large stack of wood was piled by the back door. Malcolm thought, typical white topper place. A century ago, settlers from this sect had traveled out to Nebraska to start a new colony of severely prescribed living. The venture failed, and now the remaining 'Nebraskans' were reduced to a couple of hundred families in Galloway County and a county to the north. Locals identified them by the white-topped buggies they drove, with only oil lamps hanging from either side to mark their rear end, unlike battery-run flashers on John Y.'s buggies. Henry had once remarked this branch of the Amish lived in poverty; they just didn't know it.

Jonas P. Yoder's daughter, her bare feet unwavering in the gravel-laced barnyard, laid a quart of strawberries and a dozen brown eggs on Lucille's hood. There were those in the valley, especially the white toppers, who believed brown eggs were more nutritious than white. Malcolm doubted this theory, but he liked the color and farm fresh eggs.

Malcolm reached in a jean pocket for his money clip. "*Was ist Ihr Name?*"

The girl lowered her eyes. "Rachel."

"*Danke*, Rachel." Color rose to her cheeks as Malcolm exchanged the cash to her hand. Collecting his bounty, he added, "It's little things like fresh strawberries that make the heat of summer worthwhile." Malcolm jockeyed behind the driver's seat for space while trying not to spill his berries, "Hope you still have more berries later this week." The maiden's blue eyes softened as Lucille wheeled around the barn yard.

*****

Feeling as if he had shed his skin after showering, Malcolm was pulling a carton of vanilla ice cream out of the freezer to complement his fresh strawberries when the phone rang.

"Malcolm?"

He set the ice cream onto the counter and paced the kitchen floor, his hand tethered to the wall. Stepping over Precious, who was looking intently at the abandoned carton, he braced himself for the barrage. "Hello, Mother."

43

"I have wonderful news! Your sister is going to have a baby! Imagine, Angie starting a family."

Malcolm smirked. Imagine indeed. His self-centered older sister bearing the responsibility of parenthood. There must have been a sale on chique, maternity outfits at Marshall Fields over at the nearest mall, with a matching jewelry ensemble, no doubt.

"The baby is due before Christmas, what a holiday that will be. You will come home, of course."

"I'm afraid I can't. It's my turn on emergency call this Christmas."

"Oh Malcolm, what a shame! I know you would love to meet the new arrival."

"Yes, I'm disappointed, too." Precious flicked her eyes at him and resumed interest in the ice cream.

His mother chattered on. "The Babson's daughter, Sherry, is also pregnant. She and her husband are getting a house on the other side of town. And Angie and Blake have gotten their life started, ever since they became partners in that law firm." Malcolm knew if he researched his parasitology book carefully enough, he would find a creature with his brother-in-law's name in Latin, either as a type of intestinal nematode or blood-sucking louse.

His mind tuned back to his mother's discourse. "And they have that wonderful new house, right in Hinsdale, four bathrooms, a wet bar in the den, and a hot tub. They've already selected new carpeting and curtains for the baby's room. And Blake got Angie a new Mercedes for her birthday."

Malcolm weighed in silently. What a mismatch between driving machine and skills.

His mother queried, "Are you still driving that old, um…"

"Scout."

"Yes, that's it. Isn't it time you upgraded a bit, something that a *real* doctor would drive?"

Malcolm visualized arriving at a dairy farm in a Cadillac with plush leather seats. "Kind of hard on the budget right now." His response was met with silence. The sermon is coming, which one this time?

"You know Malcolm, I think it's very nice that you want to help those farmer types, I'm sure they appreciate your work."

"On a good day, perhaps."

"You remember our family vet, Dr. Weston?"

"Uh-huh."

"He has such a nice practice. Modern new clinic with grooming and boarding, lots of people working for him, a very successful business."

"I'm sure."

"Your father often sees him at the country club, which Angie and Blake are joining by the way."

"Hope there's plenty of trees for the other golfers to take cover."

"Malcolm! Shame on you. Dr. Weston could arrange something for you at one of his other clinics."

"I'd have to find my white jacket, haven't used it since vet school."

"Honestly, Malcolm! With Angie having the baby and all, I believe you should give up that farming life and came back to start a proper, professional career. It's more of who you are. After all, look at what your sister accomplished, new home in a good neighborhood, starting a family, solid marriage…"

Malcolm smiled inwardly. At last, the crux of the sermon, entitled 'Measuring Stick of Sibling Success'.

"…and your father agrees that Angie has made all the right choices for her life."

Malcolm speculated, as well as forgetting the definition of ethics.

Silence crept into the conversation. "Well, anyway, we hope you can come home to visit, especially over the holidays."

"I'll work on it," he lied. "Say hello to Dad."

"Okay, I wanted to let you know the good news."

They said good-bye and hung-up. "Lord Almighty, they think I'm a missionary among lost souls." He regarded Lucille parked in the driveway. "If not a Cadillac, at least a car with air conditioning would be nice. He searched for a spoon and plied out the softening ice cream into a bowl. Precious followed his progress, licking her lips. Malcolm shoveled a mouthful of vanilla and strawberries. In the distance, the early evening sunlight tinted the ridges in ginger and rust. The ringing phone jolted his study of the landscape. Wondering what it would be like to be a dog and not worry about answering phones, he set the bowl on the counter.

"Doc? This is Sally. We've got a little problem that needs fixed."

"Down cow?"

"Oh no, Doc. Dale called on the radio, some sort of car wreck with a heifer. Hurt pretty bad, so she is. He says she needs to be put down. About a mile and a half up the Jericho Road, just a little before Nat Brown's place on the curve by the pasture."

"Okay, I'm on my way."

"Thanks, Doc." Malcolm recalled hearing sirens in his stream of consciousness during his mother's life coaching. Spooning three more piles of berries and cream into his mouth, he placed the bowl on the ground. "Hurry up with the dishwashing, goofball, we've got a call to make."

CHAPTER VIII

The evening shadows crept across the north mountain and pursued Malcolm along the Jericho Road. He swerved around a bend and encountered a flotilla of pick-ups with flashing red lights. A volunteer fireman waved him off the road and onto a hay field. Dale Calhoun, dressed in blue jeans and a white tee shirt, waded through the foot high alfalfa.

"Bit of a stinkin' mess, Doc. The poor thing's leg is broke really bad. Nice looking Angus, so she is."

"Who does she belong to?"

Dale shrugged.

Malcolm pointed to a knot of people about seventy feet away. "All those folks volunteer firemen, Dale?"

"Hell no, Doc! Most of them are gapers. Get 'em all the time, fires, car wrecks, severe storms and the like."

"How they get here so quick?"

"Everyone around here has got police scanners."

Malcolm gazed at Precious. "Stay here, too many people and an unpredictable patient." She flattened onto the seat and groaned. "Stop complaining."

Malcolm followed Dale, alfalfa stems brushing his calves. They filtered through the onlookers and came upon a jet black heifer. The heifer bawled and tried to butt Malcolm as he closed in. She catapulted forward, flopped onto her head and rolled onto her side. The unnatural bend in her right rear leg was all Malcolm needed to sum up the situation. Pity and grief swept within Malcolm. "Judas Priest."

Dale said, "I guess the family in that old Plymouth was coming around the curve there and found her wandering on the middle of the road, right in damn front of them. They jammed the brakes, but the crazy thing actually charged the car. They couldn't help but hit her. She must weigh all of 900 pounds. Did some damage to the car, so she did. It needs towed. No one badly hurt though. It's a hell of a thing to be driving along and run into something like that."

Malcolm thought, hell of a thing for the poor heifer as well. Her alert eyes scanned the two-legged creatures surrounding her.

"Watch her, Doc, she's kind of ornery."

"I'd be out of sorts too, if my leg looked like that." He raised his voice. "Does anyone know who owns this animal?" A chorus of negative mumbles responded. "The fracture is beyond repair and she can't manage to get up. Without knowing the owner, I've got to euthanize her." He was answered with silence.

Malcolm guessed the heifer's weight—her anxiety level was through the roof—he wanted to get the drug dose right. Turning about, he commented, "I'll get my things from the Scout, Dale."

An older man with a straw hat peeled off from the circle and limped next to Malcolm. The visor of the hat and stooped posture completely obscured his eyes. "You know young fella, that heifer is worth a fair amount for beef. She shouldn't be wasted like that."

Calculating the lethal dose of barbiturate in his head, Malcolm replied, "Sorry, but the euthanasia drug makes her unfit for consumption."

"Why don't you get the officer to shoot her, someone could use her then."

Malcolm considered the ring of spectators. "Safer and more humane for her to use the drug."

"We could reach some sort of arrangement," the hat persisted, "It would work out for both of us."

Malcolm disliked the direction of the conversation. "No permission from the owner, no slaughter for meat. I don't want to find out later I gave someone's animal away to the wrong person. Besides, I think she's suffered enough."

The hat stopped short of the Scout. "Well, that's a sinful waste of good meat so it is. A sinful waste!"

"Sorry, it's the right thing to do, for her sake." Malcolm opened a back door in Lucille.

The hat stepped forward. "You look here…" Precious thrust her head out the passenger window and snarled with back hair standing on end. The hat retreated to the assembly of observers, who shifted their interest from injured animal to the conflict by the Scout.

Malcolm stroked his companion. "I don't like him either."

Dale sauntered over with a police officer and a fire company buddy. With a glint in his eye, he said, "He's pretty pissed at you, Doc." Malcolm looked up from filling a syringe, remained silent. "You know why he's pissed don't you?" Malcolm knew if he waited, he would hear the answer to the question. Patience wasn't one of Dale's character strengths. "That's old Haumiller, he's got the custom locker plant in town."

"You mean that little place by the creek?"

"Yeah, that one."

"And so?"

"Folks here saying the heifer broke away from his lot the night before she was to be killed. Cattle break through his fence all the time. She got up in the hills some two months ago and has been roaming ever since. People seen her about, running about pastures and corn fields like a wild antelope."

"So what's the deal, Dale?"

Dale relished the elucidation of the obvious to the well-educated, although naive-to-local-affairs, veterinary. "He was probably going to custom kill her for somebody, their own backyard beef animal. Old Haumiller must have come up with another animal to replace her. Folks probably never knew whose beef they were eating."

"Well if the heifer is his, why not just say so?"

"That old piss pot won't do that. State law says he'd be responsible for the car." The officer nodded. "She's got no ear tags or nothin', can't prove it's his. He don't want to pay for the front end of that car."

"So what does he want?"

"He wants you to let her be, let Tommy put a bullet in her, and then he'll volunteer to take her to the slaughter house." Dale stole a quick look over his shoulder. "But you got him by the short hairs, Doc. You put that poor thing out of her misery. If he can't 'fess up to the wreck, he don't deserve her. He won't start nothin' 'cause we've got you covered."

Malcolm rolled his eyes. Justice in Galloway County, twisted as it may be. He said, "It's the humane thing to do." He sorted out a lariat and halter from underneath his calving gear and returned to his mission with his posse tailing.

The hat spoke tersely, "I can't believe a so-called doctor would just go and waste a good animal like this." He beseeched the gathering, "Youens all know there are people that could use that meat, don't you know." A whiff of discontent grew in the fading sunlight.

Malcolm raised his voice. "No owner, drugs go in animal, no slaughter." Dale simpered as the corners of his mouth lifted. Haumiller kept muttering as Malcolm approached the heifer, lariat in hand.

The heifer bawled again, her shattered leg dragging on the ground as she crawled forward. Malcolm got the lariat about her neck on the second try. She snorted and pulled against the rope, violently tossing her head and lurching at Malcolm. Dale pounced to add his weight to Malcolm's, his strong arms animated like an angler with a big catch. Malcolm got behind the heifer and jabbed the tranquilizer into her hind leg. She bellowed as Dale held onto the rope for dear life.

Malcolm hand-signed Dale to ease up. "She's awfully aggressive, must be in unbearable pain."

"Or just being a typical Angus," Dale said.

Malcolm wondered about the farmer's explanation, let alone road rage with the car. He wished the gapers would leave. The heifer didn't need them to add to her apprehension. Malcolm kept his distance from her, waiting for the tranquilizer to spread tendrils of calm into her mind. His mind wandered to Haumiller and his crappy fence. *Maybe I'm being peevish to waste her life. Too late now, she's had the shot.*

The sedative prevailed. The bovine head sagged and she sang her death song in a relaxed, low moan. Malcolm approached her with the halter and slipped the rope behind her ears despite her feeble attempts to resist. He squatted to find the jugular. "No more pain you poor thing, go to sleep," Malcolm whispered as a red cloud of blood mushroomed into the syringe and surged back into the vein with the barbiturate. She gasped once and collapsed. Malcolm sighed as he looked into her glassy eyes.

He stood back and stared at his effort with emptiness. The sun was sinking into the horizon, a somber red disc in the haze. The spectators dispersed.

Dale joined him. "She can't go for meat, can she Doc?"

Malcolm let out a breath. "No way, Dale, not with the drugs I gave her."

"Well, we can get the renderer to come pick her up and dispose of the body, but it'll probably take an hour or so to get here."

"Gilkey's will come out on an evening call?"

"Yeah, if the cops ask 'em."

Malcolm nodded. This county was full of people who always seemed to work long hours for little pay, some days he figured he was part of that parade. "Good, that's where she should go."

Dale said, "Pretty bad stuff then, if that heifer gets eaten?"

"Could be toxic, yes."

"Well then, I think me and a couple of the boys will stay here with Tommy until Gilkey's gets here to watch things."

"What things?"

"Christ Almighty Doc, you can bet if we didn't stay, ol' Haumiller would be back with a couple of his gooneys and pick her up."

Malcolm stared at Dale. "What kind of person would do that?"

Dale's eyes narrowed as he looked over Malcolm's shoulder. "He's kin folk to MacDougall, don't you know." He added emphatically, "Once't we all leave, she would too."

"Well, if you think so, Dale."

"She needs watched, so she does."

Noting Haumiller conversing with someone in a pickup by the road, Malcolm retreated to Lucille, eager to leave the scene. As Lucille wheeled about the hayfield, Precious growled when the headlights flashed across her adversary.

Haumiller yelled out, "You call yourself a vet?"

Precious answered with a volley of barks out the window. Thinking of Haumiller's familial connection, Malcolm fetched her back inside by her collar. "Let's not provoke unnecessary quarrels."

CHAPTER IX

Long days flowed by in a narcotic current of heat and humidity. It was the time when gangs of men with bronzed arms and perspiration-stained white shirts stacked shocks of wheat under a hostile sun. Barn odors and sweat oozed from long-haired scalps and brows and stung the eyes. After a cool shower that washed away the toil of yet another searing day, Malcolm studied the rough-cut timber sign over the door, **Dead Bear Bar and Grill**. The letter "r" was missing from "Grill" and a window in the deserted second story wanted for a pane of glass. This establishment and the fire hall were the social hubs in the village of Salty Dog. Malcolm mulled over the clientele and activity that lurked within. Nonetheless, neon Budweiser and Rolling Rock signs in the front windows glowed invitingly and aware he had been drinking by himself a little too much lately, he took a deep breath and entered.

Bob Seger drifted from a jukebox across a miasma of stale beer and cigarette smoke. A dull cue ball dispatched its solid-colored target over the threadbare felt and into a pocket of the pool table. Malcolm got a Yuengling from the bar. His eyes adjusted to the dim light and found Bobby, who was watching a heavy-set man fling red-feathered darts at a board on the far wall.

"So what are the rules?"

"Each team starts with 301 points and you subtract the points you get in each three-dart toss until you reach zero. But you have to reach exactly zero, or your busted back to where you started that round."

"Who's winning?"

"I think we're both near a hundred. Progress slows down as you get near zero, though." Bobby put his beer down. "Time for another toss."

Malcolm tried to stay tuned into the match, but he got another beer, took a leak and fantasized on the potential reception he might get from a red head on the other side of the bar, until her boyfriend arrived. The accuracy of dart placement waned considerably as the evening progressed. Finally, the local team whooped and cheered, the apparent victors.

"I'm swollen with Salty Dog hometown pride", Malcolm said.

Bobby rationalized, "It's just as good we lost, keeps the likelihood of fights to a minimum."

"Over darts?"

"Some folks take the outcome pretty seriously."

"Fueled no doubt by equally serious beer drinking."

Chaotic discourse broke out across the room. A earnest, marginally sober trio split off from the group and converged on Malcolm and Bobby's table. Bobby said, "Looks like you got yourself a welcome committee."

The well-groomed leader of the delegation strode on lanky legs, squeezed into tight jeans fixed to the waist by a silver belt buckle in the shape of a horseshoe. The second envoy wore a tee shirt and jeans, and the third, a bowling shirt and polyester pants. The ample forearms of the fellow with the tee shirt cradled a bewildered small white and tan dog.

Malcolm recognized the one in the western wear as Shannon McMorrow, the town barber and a most amicable drunk. Startling blue eyes radiated under his thick jet black hair while his smooth voice implored, "Doc, we have a predicament here. We can't settle on the family heritage of this puppy." Malcolm considered Shannon's alcohol-soaked companions; their grins revealed logic was currently off their radar.

"Do you mean you're trying to figure out the breed of this dog?"

"Exactly, Doc! You see Jimmie here," Shannon gestured to the tee shirt, "Says he's part Brittany Spaniel, but Ronnie keeps carrying on about that he's a beagle with long hair."

The presumed bowler peeped through dark-rimmed glasses. "Put him near a rabbit and he'd run clean on into the hills."

"This ain't no goddam hound, he's a bird dog," Shannon retorted. Turning to Malcolm with a smile that could thaw snow on a cold January day, he added, "I declare there's a fair amount of wagering on this Doc, if youens want a piece of it." Across the table, Bobby suppressed a smile behind his hand.

Malcolm rubbed his chin. "That would be a slight conflict of interest on my part, Shannon."

Shannon placed his right hand over his heart. "I know you'll be an impartial judge of unquestioned integrity."

Malcolm was overwhelmed by the trio's confidence in his professional ethics. "What's his name?"

"Tadpole. He's Cam Shoat's dog."

"Okay, let's see Tadpole." The timid canine appeared on Malcolm's lap, wagged his tail, and raised forlorn eyes. Malcolm pet his head and empathized. *What are you doing in a place like this entertaining these clowns?* Running his hands gently over the back, legs, and rump, picking up the ears and lowering the eyelids for good effect, Malcolm pronounced with gravity. "In my professional opinion, Tadpole is a mix of Basset and miniature Irish Setter. I'm afraid your opinions are a draw, gentlemen." He carefully transferred Tadpole back to Jimmie.

Shannon guffawed, "Son of a bitch! I knew it! What did I tell you, part bird dog with the setter in him."

Ronnie said, "To hell with that! Didn't you hear? He said he's part Basset, that's a hound dog, like a beagle." Arguing, the committee returned to the far corner to relay the judgment. Malcolm watched Ronnie swagger across the bar, any number of species of poultry could claim less scrawny legs.

Bobby smothered a smile. "What a clown show. Miniature Irish Setter?"

Malcolm shrugged. "This will keep them from losing all their money before going home to their wives and kids tonight."

"There was already considerable changing of cash on the darts game."

"So *that's* what makes the outcome so important."

Bobby looked over Malcolm's shoulder. "I think that gal tending the bar has been checking you out."

Malcolm jabbed at Bobby with his index finger. "Don't start."

"I could probably break the ice for you."

Malcolm stretched his back. "Time to go."

"Yeah, I guess I do have milking duty tomorrow bright and early…chicken shit."

On their way out, they were enthusiastically saluted by Shannon and his cohorts. The blond behind the bar winked at the veterinary hero. Malcolm smiled in return; he had performed his good deed for the day.

CHAPTER X

The next day stretched into another long, hot trek in the valley. Having returned from a difficult calving, Malcolm thought his day might finally be over. The answering machine proved him wrong. "Hulloo, *junge Herr Tierarzt*. John Y. Peachey here. I have a bred heifer with something with her walking that's wrong. Maybe you should see her this evening, please." Malcolm's stomach growled. He hadn't eaten anything since the morning coffee and doughnut in the office. He slapped peanut butter and strawberry jam between two slices of bread. "Come on, jellybean." Precious dashed out the door, while Malcolm tried to pull the handle without smearing excess jam over the knob.

The cows were already out to pasture for the night, leaving John Y. behind to clean up the empty stable. He paused to lean his arms on the handle. *"Guten abend, Herr Tierarzt."*

"Problem with a heifer, John?"

Scratching his chin through his thick beard, the farmer answered, "Yesterday evening I sent the boys to get her with the others and she acted like she didn't know where she was going. Kind of like you finding your way around the valley."

The corners of Malcolm's mouth creased upwards. "Going where, John?"

John Y. looked surprised. "Why, out of the pasture of course, she's springing."

"She's due to calve soon?"

"*Ja*, I think so."

"So she had trouble making her way to the barn?"

"*Ja*. Kind of stumbling, had to push her to keep her going, so the boys did. Didn't want to stay with the others."

Malcolm pouted. "That's odd. Where is she, John?"

"Out behind the barn, come with me." John Y. pointed to Precious, "And bring your help." The farmer shoved aside a large sliding door in the back of the stable with ease. "*Komme, Tierarzt.*"

Barn swallows chattered above their heads, sweeping up their last meal of the day. Farmer, veterinarian, and dog converged on a large shed, enclosed on three sides, with a fence across the entrance. "Stay there," Malcolm directed Precious as he climbed over the rails. Two heifers got to their feet and tramped in the loose straw to a far corner of the pen, but the third herd mate lacked concern over the human intruder. Her head drooped below her shoulders as she wandered about the pen with little sense of purpose.

"Has she gotten worse since you brought her in last night?"

"*Ja*, I think so."

Malcolm sized up restraint for his patient. "John, could you please get a gate tied up in the corner of the pen? I'd like to examine her a little more closely."

"*Kein problem.*" John Y. strode around the corner of the barn. Malcolm relished the approaching sunset and the cooler air it would bring. Wisps of orange-tinged fleece dappled the washed blue canvas on the horizon while a mourning dove crooned a melancholic ballad from a spruce tree by the side of the barn. John Y. reappeared, escorted by his two sons, who carried a ten foot steel gate between them. The farmer led his sons to a corner of the pen and tied one end of the gate to a post with baling twine. Farmer and sons circled the heifer and coerced her into the makeshift crowd gate.

Finished with his physical exam, Malcolm shoved his stethoscope in a back pocket. "She has a fever, John, although not too high considering the warm weather. When is she due?"

"Hard to say, bull breeds them back in the winter." The farmer's eyes twinkled. "He enjoys his visits on long, cold nights with his girlfriends, so he does."

Malcolm smiled. "Not unlike farm boys." John Y.'s sons

lowered their heads and smirked. With daylight waning, a line of shadow masked the heifer's face. Bending lower, Malcolm peered more closely at her eyes. "Bit of a puzzle, John. I'm going to get a couple of shots from my car."

"Better get some smart pills, veterinary, she's a little bit *ein dumm Esel.*"

Malcolm scaled the fence and scuffed his boots in the loose dirt around the barn. A little bit of a dumb ass indeed. Precious drubbed her tail on Lucille's rear panel as he opened the hatch. "No, we're not going yet." Malcolm rooted about the paraphernalia in a large box. He grabbed a magnet on the off-chance the heifer had eaten a piece of wire. What sort of animal would eat pieces of barbed wire? He bounced John Y.'s comment in his mind. *Dumm Esel* heifer—dumb ass heifer.

His thoughts shattered like a stone through glass. "Son of a bitch!" Grabbing a flashlight, he tramped back to the shed. Precious trotted to maintain pace. Long day or not, he could kick himself as he put the pieces together; uncoordinated gait, dumb attitude, and depression. "Damn it!"

John Y. and his sons glanced at one another as Malcolm marched straight to the heifer and aimed the flashlight at her face. The beam of light constricted both pupils in their turn. The right upper eyelid appeared a little droopy.

"John, what have the heifers been fed this past week?"

John Y. raised one eyebrow. "Why, just the pasture."

Malcolm assayed the other heifers. "Give them any medicines?"

"No."

"Dump anything in the pasture?" John Y. shook his head. "John, it's important to keep her somewhere away from the other animals and for you to keep away, too. Put out some hay and water for her, then stay away. I'll give her some antibiotics, in case it's meningitis. I'll be back tomorrow morning. Can you rig up a separate pen for her with some fence gates and the like?"

The farmer nodded and looked at his sons. "*Ja.*"

Malcolm stuffed his gear into Lucille, while Precious launched herself onto the seat. He sloshed soapy water out of his bucket with the brush and banged his boots with a purpose.

Precious pushed her head out of the passenger window. "R-r-r-ruff!" Malcolm finished cleaning to find John Y. carrying a plate.

Malcolm said, "Keep on bringing treats and she'll end up wanting to stay here."

"A good *Hund* leave her master for a moon pie? *Nie!*" John Y. extended the plate to Malcolm.

The pastry warmed Malcolm's palm as he chomped a mouthful of mincemeat. "*Danke*, John, this is good." A whimper rose from behind. Malcolm broke off a morsel that vanished from his hand.

John Y. laughed. "Now who spoils her?" Chewing on his own moon pie, the farmer asked, "What ails the heifer?"

"Can't be sure yet, John, I'll wait until we see her in the morning. I'll stop by before I go into the office."

"*Gute Nacht*, veterinary."

"In the meantime, keep your distance."

Reaching the end of John Y.'s farm lane, Malcolm regarded the yellow moon hanging over the shoulder of the south ridge. Below, rank and files of wheat shocks lined the fields; hooded specters advancing to their doom on the thrashing floor. Fireflies flickered among the straw recruits in search of their dancing partners. Sharing his moon pie with Precious, Malcolm contemplated John Y.'s heifer. Overhead, a small creature gyrated in erratic flight. "They're out already, barn swallows by day, bats by night."

The sultry air still hung in the house as Malcolm checked the answering machine, no more calls. He fell into a patio chair and swirled three fingers of Oban around the small jelly glass. Creamy lilies and daisies preened in the opaque moonlight of his garden. Immersed in the aesthetic sparseness, he could often distance himself from the clutter of his mind. But he failed to drift away into oblivion this evening, uneasy over two heifers, the one with the vacuous eyes at John Y.'s and the Angus that jousted with the car.

CHAPTER XI

A scarlet vein kindled the eastern horizon. The radio droned on with an unnecessary forecast—the heavy air foretold another day of bone weary heat. Malcolm chased a couple of sunny sides around a skillet sizzling with butter. Precious devoted her attention to the buttered toast on a plate. Malcolm loaded the eggs next to the toast and sat at the mail-littered table. "Stop begging," he scolded Precious as a piece of egg on toast vanished under the table. Sipping his coffee, he flipped the pages of a textbook to neurologic disorders of cattle. Satisfied with his research, he placed the plate on the floor for canine cleaning and rinsed his coffee mug in the sink. It was time to drop by John Y.'s on the way to the office.

Early morning light at his back, Malcolm leaned over the fence of the makeshift pen, bolstered with a manure wagon and baling twine. The heifer loitered about with her nose nearly touching the ground. Malcolm whistled in a sharp pitch. She wavered on her feet and lifted a listless head in his direction, her tongue extended. Losing interest, the heifer sank back into her own space. Malcolm found a piece of two by four leaning on the barn wall and eased over the fence. He approached her from the rear and meticulously tapped her rump with the end of the board. Her head quivered with a subtle tempo. He shoved the end of the stud into her stifle. She staggered as if rolling out of a bar at closing time.

Malcolm sought out John Y., milking in the stable. "John, I want to get Henry's thoughts on the heifer, maybe come back and see her later today."

John Y. rubbed his neck. "Something serious, *Tierarzt*?"

Malcolm bit his lower lip. "I'm afraid so. Did she eat or drink?"

"I don't think so."

"Not even a little water?"

"I would have to look in the pail we left her last night."

Heading for the open stable door, Malcolm called out, "I'll be back…and stay away from her!"

Chasing after Precious into the office, Malcolm nearly bumped into Henry with two mugs in his hand. Lois was stocking shelves. Malcolm tried to restrain the excitement in his voice. "I think I have a real case, a cow with cortical signs, strange behavior, strange, *dumb* behavior."

Henry said, "A real case of what?"

"Rabies! At John Y. Peachey's! A heifer that came out of the pasture. I saw her again this morning. Who do we call to report this?"

Lois turned her head. "Well, we better call the state vet's office, the *News-Gazette*, AP wire service…"

Henry grinned. "Relax, collect yourself."

Lois grabbed the mugs from Henry and offered one to Malcolm. "Have a cup of coffee and tell us the details." Henry contemplated his now empty hands and retired to the back room. Malcolm described the progression of the heifer's symptoms.

Returning with a fresh mug, Henry asked, "The dumb form then?"

"Unresponsive and unaware."

Henry pursed his lips and clasped his coffee with both hands. "Maybe you should check her later today."

Lois offered a biscuit to Precious. "Henry, you should go, that animal could be dangerous."

Malcolm said, "I'll take care of it."

Lois suspended the next treat about twelve inches above Precious' frustrated eyes. "We'll set things up so that both of you can meet at John Y.'s at the end of today's calls."

Henry swallowed a slug of coffee. "If we suspect rabies, we'll want to call the Suumerdale diagnostic lab to let them know we'll be bringing the head."

The much anticipated biscuit finished the journey to Precious' mouth. "I think Precious should keep me company today." Lois produced two strips of adding machine paper scrawled with the call lists. "You're in for a long day. You guys better get moving if you're going to meet at John Y.'s"

*****

The sun bore down from a cloudless sky.

Malcolm shuffled his way to the makeshift pen. "What do you think Henry? Was I right?"

"Well..." Henry massaged his jowls.

The heifer struggled to maintain her balance, each respiration ended in a groan. Slight tremors vibrated her head from side to side, a pendulum marking her waning minutes on earth. With a large thump, she sank to her knees. A cloud of small black flies swarmed away from her in a buzzing array. They returned, crawling about her face, back and legs. The heifer moaned pitifully.

Malcolm couldn't pry his eyes away. There had been movies in vet school; scenes captured on grainy film of a horse or goat in a distant third world country, prodded by men in local attire, trying to demonstrate the clinical signs of rabies. The gravity of the situation was hard to appreciate, especially in the comfort of a lecture hall ensconced in the tedious daze of six or seven hours of lecture a day. In his mind, rabies was an exotic disease of textbooks and remote places.

Henry ended Malcolm's hypnosis. "I'm afraid your concern over rabies may be genuine, Malcolm."

"I'm surprised how fast she's going."

"We need to put her down. This will be tricky. We can't halter her and administer the IV. We'll have to shoot her, but not in her head, need that for the lab. John Y. should be here soon, I waved at him and his boys in the hay field. We can probably get him to use his gun."

63

They waited, with the agonizing heifer, the horde of flies, the heat, and the sweat. Even the shade seemed oppressive. The clothes beneath Malcolm's coveralls clung to his body. Some of the flies bit at his face and ears.

Henry remarked wryly. "You smell rather ripe today."

"Yeah, had a retained placenta at Louis S. Yoder's."

Before Henry could reply, the sound of horse drawn equipment clattered near the barn. John Y. came around the corner, adorned with flecks of hay dust on his damp neck and face.

John Y. nodded to the onlookers. "Why, such a case is this to need two *Tierärtze*."

"Hello, John, this is not a good situation for you or your heifer."

John Y. looked at the heifer. "Ja, she's in a bad way I'm afraid. What is it Mr. Breck?"

"I think Malcolm is right, perhaps rabies."

John Y. turned his attention to Malcolm, then Henry. "Rabies?"

"We might need your deer gun, in case we can't give her the drugs to put her down."

John Y. lifted his straw hat, pulled a blue bandana from his pants pocket and wiped his forehead under a ring of matted hair. He replaced the hat and said, "Well, this is something," and disappeared into the barn.

Malcolm observed the facial paralysis on the heifer. The flies took advantage of her inability to flick her ears, miniature black harpies bent on an orgy of greed…damn flies. "Henry, I'm going to go get the drugs, some sleeves, a mask, and my hunting knife."

"She might be as likely to jump up as not when you stick her with the needle. Things could get ugly if you get knocked down."

"I have to try."

Sweat trickled down Malcolm's neck and temples as he traversed to and from Lucille in the unforgiving sun. Malcolm was joined by John Y. as he exited the side door of the stable, escorted by his sons. A .22 rested in the farmer's arms. Malcolm inquired, "John, would you happen to have a couple of plastic feed bags? I could use those for carrying the head."

John Y. stretched his head back over his shoulder, "Eli, *bitte bring zwei Futtersäcke für den Tierarzt.*" The younger son pivoted and faded into the barn.

The heifer remained on her knees, moaning and bobbing her head. Malcolm edged over the fence, his stomach in a tight coil as he gauged the twenty feet to his subject. "Careful Malcolm," Henry counseled. Malcolm approached the heifer from behind, an overdose of tranquilizer in his right hand. His quarry seemed oblivious to his presence. He pulled the cap off the needle and jabbed the syringe into her rump. The heifer bawled and stumbled forward. The needle hub snapped off the syringe and some of the drug sprayed over her back. Malcolm grimaced. "Son of a bitch!"

"Stay clear of her, Malcolm," Henry said.

Human and animal alike bided their time while the flies buzzed on. Malcolm fidgeted, unsure of how much tranquilizer was injected before the syringe snapped. "Henry, I wonder if I should pull another syringe?"

"Give it a minute or two, I'd rather you stay away from her."

With a groan, the heifer managed to turn and face Malcolm, her feet shifting without synchrony. Malcolm observed with relief the copious amounts of saliva stringing from the mouth. He turned to give Henry the 'thumbs up' when Henry's eyes widened. "Malcolm!"

A deep rasp shot from the heifer's throat as she bounded towards him. Malcolm angled backwards and brushed off her right shoulder. She fell onto her left side as her legs swept wildly in an arc. Driven off balance by the heifer's mass, Malcolm twisted to his right to avoid the more dangerous hind limbs, but tripped over a front leg that clipped his knee cap. He rolled to his right, scudded along the concrete and onto his rear end. Rough hands grabbed his armpits as he pushed to gain his feet. Henry said, "Not a good place to be."

Malcolm gathered his wits. "Thanks Henry, I'm okay."

The heifer's respirations diminished into low hums. "I think we're getting where we need to be," Henry said, lifting a syringe full of blue fluid from his coverall pocket. The heifer progressed into slumber, her breathing shallow and irregular. Malcolm said, "Please Henry, let me finish this."

Henry analyzed the heifer's demeanor and handed him the syringe. "Go for cardiac puncture, I don't trust hitting the vein."

Malcolm circled the debilitated beast, the heat and the enclosure all too similar to an arena. He leaned over her back, palpated a space between the ribs with his fingers and pushed the plunger. Her erratic breathing stopped instantaneously. Malcolm breathed a sigh of relief, her agony had ended.

Wearing a mask and two sleeves over each arm, Malcolm unsheathed his knife. The six inch blade severed the throat and blood flowed onto the concrete, a river of misery that attracted the attention of the infernal flies. He longed for a canister of spray, to avenge their torture of the heifer. Pushing her head towards her chest with a boot, Malcolm continued slicing across the back of the neck, and with a final cut of the spinal cord, the head was free.

Malcolm set the knife down, got a feedbag from Eli, and returned to the severed head. He lugged his ponderous trophy to the fence and placed it into another bag that Henry held open. I'll go get a bucket for you," Henry said as he hauled the bag towards their vehicles. Malcolm remained, not wanting to touch anything, sleeved hands in the air.

Henry returned, a jug of ninety-nine percent isopropyl alcohol in one hand, a stainless-steel bucket in the other. He poured the entire jug into the bucket, suggesting, "Malcolm, place the knife in here and let it soak a while." Malcolm walked across the dusty pen and reluctantly picked up the knife, a murderer forced to retrieve his weapon. Steel scraped on steel as he gently slid the knife into the bucket through the fence. Henry had brought another plastic bag, holding it open while Malcolm dropped the syringes and stripped off his sleeves, followed by his mask.

The clank and rattle of steel and iron heralded the arrival of two sturdy Belgians, dragging a heavy chain. Samuel held the reins, walking behind the team. With a low "Whoa", and a gentle pull, the team stopped. John Y. cut the baling twine that held the fence together, swung it open and directed his son to get the team in place.

The heavy-footed beasts lurched into the pen. Samuel pulled the team into a turn and stopped. With steady palms on the chests of the giants, their blond manes towering over his hat, John Y. spoke softly, "*Zurück, rückwärts,* Pete." The horses stepped back, one hesitant step at a time, nostrils flaring from the scent of blood.

Malcolm lifted each rear leg of the heifer to wrap the chain around the hocks.

"John, this animal should probably be buried so dogs, coyotes, and the like can't get into it," Henry instructed.

John Y. nodded. "*Ja, Herr Breck.*" John Y. turned to Eli. "*Schaufeln bitte,* Eli."

"Wie viele?"

"*Drei.*" John Y. smiled. "Four, if the young *Tierarzt* would like to help." Eli disappeared. John Y. directed Samuel to take the team out to an open area away from the barn. With a whistle and a flick of the wrists, the team drove their hind legs, Samuel walking to the side. Bouncing like a rag doll, the heifer's carcass brought up the rear, leaving a trail of flattened weeds.

Henry poured some of the alcohol onto the ground where the heifer had bled out the last of her life. "Can't be sure if some of the barn cats or the dog might get at this."

Malcolm stared at the rusty spot as Henry ignited it with a match. Blue flames fanned across the ground. Eli walked past the pen, the light of the fire glinting off three shovels lying across his shoulder. Henry stated in a flat tone as he watched the blaze, "John, when you push the heifer into the hole, don't go near the neck. Also, make sure you wash your hands well when you are done. I know this all seems a bit much, but animals, including people, rarely survive rabies."

"We'll do that, Mr. Breck." The fire dwindled; signaling that the exorcism of the heifer's demons was complete.

Malcolm said, "Well, I guess I've got a trip to the diagnostic lab this afternoon."

"I radioed Lois to let them know you're coming."

Malcolm picked up the pail with the knife, wished John Y. luck with his tasks and walked back to the front of the barn.

Henry watched Malcolm, stroking the knife with furious sweeps of the brush. "I'll take your knife back and autoclave it." Malcolm handed the handle and sheath to him. The bagged head was on the floor in front of the passenger seat of the Scout.

Henry looked at his watch. "You should have plenty of time, make sure to explain the circumstances of exposure when you get to there. Lois said she would drop Precious off at your home."

Malcolm started the engine. Raising his hand he called out, "I hope we don't have to go through this again." Lucille circled and ascended the dusty lane, leaving Henry standing in the sun drenched barn yard. "A little disarrayed, yes," Henry thought to himself, "but not afraid to jump into the middle of things." Henry removed his cap and rubbed his glistening scalp. His fingers stroked the old scar, compliments of an angry Black Angus in a chute and jettisoned nose lead. His wife was still alive back then. She had always cared for him and the invariable aches and pains of his daily labors. Henry rehashed the choices for John Y.'s heifer. Over twenty years in practice and he had never seen anything like this. Malcolm's vehicle skidded from the farm lane and onto the road, the sound resonating across the hay field. Henry wondered if that old Scout would last another year given his protégé's driving style. Was I that reckless when I was his age?

*****

Three days later, Henry stood in the office and waited for Malcolm to pick up his ringing phone. "Malcolm? This is Henry."

"I thought you were a farmer ringing me for an emergency."

"No, hopefully things remain quiet this evening. The diagnostic lab called. John Y.'s cow was positive for rabies."

"Where do we go from here?"

"Not sure, probably a result of an epidemic of rabid raccoons in the state this year."

"I guess I'll add this to my list of life's experiences."

"Just wanted to let you know."

"Thanks, Henry. I'll stop by and tell John Y. on the way into the office tomorrow morning. See you then."

"Good night, Malcolm." The line clicked dead on the other end.

Lois ceased her shelving of penicillin bottles into the refrigerator. "That boy has the ability to get in more fixes than a nosy tomcat."

Tapping his fingers on the phone receiver, Henry replied distractedly, "I hope none more troublesome than John Y.'s heifer."

Lois drew near to Henry's broad back. Laying her hands on his shoulders and rubbing the base of his neck, she whispered, "Wanna bet?"

*****

The day had been longer than most and Malcolm felt like a rusty spring pulled from either end by a couple of locomotives. He pondered Henry's news, wondering if rabies would spread about the valley to other farms. The phone rang. He took a hard look at the distraction, thinking *the Amish are fortunate to leave phones out of their lives.* Being on call, he guessed the caller's identity and what the problem may be from a list of possibilities; none of which appealed to him as tired as he was. Picking up the receiver, Malcolm was relieved and surprised to find out how far off his guess had been.

Denied the sound of her voice for nearly a year, he was unable to cross the void and respond to her greeting. She probed again, "Malcolm, are you there?"

"Yes, Carrie, I'm here."

"Are you doing okay?"

"Yeah, I'm alright, it's been a long day. How are things with you?"

"I'm doing well. How's life with the cows?"

"Still black and white and giving milk."

She softened her tone and her voice imperceptibly warmed his empty house. "That seemed to be the norm there."

"It's good to hear your voice. I've been wondering what you..."

She intervened, "Malcolm, I hope you're doing well, but we need to talk about something." His shoulders sagged. "I had to contact you regarding some of the attorney fees."

"From the divorce?"

"Yes, it's finalized now."

He put his face into his hand and sat down in a kitchen chair. "Carrie, are you sure about this?"

"Malcolm, it's already finished, we can't go back." He didn't answer. "Malcolm, are you still there?"

He mumbled, "Yeah."

"This is hard to do, but we have to go through with this." He continued to listen in silence, his temples throbbing as she spoke. "Your share of the attorney fees is about seven-hundred and fifty dollars."

"Geez, Carrie, that's nearly two weeks of my take-home pay."

"That's part of our settlement, shared expenses."

"How did the bill get to be so much?"

"There were court filings, letters, and things. The attorneys are bothering me that you haven't responded to their mail."

Malcolm looked at the clutter of unopened mail on the counter. "I guess I'll get on it."

Her voiced ramped up. "Soon, Malcolm."

"Well, why did you have to file in Illinois?"

"Because I set up my legal residence here before filing."

"That was clever."

The intensity of her voice escalated. "My attorney advised me to do it."

"Swell."

"You don't have to be a smart ass. If you'd give up your taste for expensive single malt, you might be able to afford to pay your bills."

Malcolm concentrated on keeping his mouth shut. After an awkward silence, she asked, "Have you been out to your parents lately?"

"No, not sure when I'll fit that in." He proposed an option. "Perhaps if I did, I could stop by to say hello."

"Not a good idea, Malcolm."

"Why?"

After a pause, "I'm seeing someone."

"Who is he?"

"Does it really matter?"

"What does he do?"

"If you must know, he works at my father's firm."

"No doubt he has the same endearing qualities as your dad."

Her response was slow in coming, but her voice was clear and cold. "I'm engaged Malcolm. He is a more successful man than you could ever hope to be."

Her words spilt open his skull. Without thinking, he lashed out, "You mean to tell me you were engaged before the divorce was final? What kind of bullshit is that?"

"Just pay the attorney bill, Malcolm."

Malcolm ranted, "Carrie, how could you betray me like this? I don't hear from you all this time except when..." The line clicked dead, she was gone. He held the receiver in his hand, studying the beige plastic as if it was a unique artifact.

He located the recently maligned single malt in the cabinet above the refrigerator. Like an impressionist painter, he needed to soften the edges of the day's canvas.

CHAPTER XII

A cold nose in an ear rattled Malcolm from deep slumber. His mouth tasted like dry rubber. "What do you want?"

"Woof!"

Malcolm's left eye slit open to see a pair of brown eyes six inches from his face. Stiff neck muscles complained as he twisted his head on the sofa cushion to confront his tormentor. He came to grips with the reality that he had failed to stumble into his bed last night.

"Woof!"

"I suppose you have to go out." He propelled himself onto unsteady legs and checked the clock on the wall. Damn it! I've got to move, no time for a shower. Pressing Precious out the back door, he searched for a pair of jeans and shirt with the most neutral odors from his laundry pile. He brushed his teeth and splashed water across his face, then urged Precious not to dally in the back yard. Grabbing the keys for the Scout from the kitchen table, he rushed out the front door.

Lois was sorting out bills at the desk when Malcolm plodded into the office. Gauging the entire length of his body, she said, "Let me guess, you chose to sleep in a neighbor's doghouse last night."

Malcolm scratched his unshaven face. "In a fashion, yes."

Henry appeared from the back room, coffee mug and doughnut in hand. "We've got doughnuts, Malcolm."

"No doughnut yet, just coffee."

Lois arched her brow. "Rough night?"

"More like long day." Malcolm shuffled to the back room, poured coffee into a mug and seared his mouth on an eager gulp.

Re-entering the office, he found the bench along the wall to have more than the usual appeal.

Tapping a pen on the desk, Lois asked, "Henry, have you finished your invoices from last week?"

"I was out bowling last night, Lois."

Rising from the desk, she poked her pen at a wall calendar with placid grazing Jerseys. "Huh. My calendar tells me that today is Thursday, and we all know that it is company policy that weekly invoices are due on Thursdays." She swept towards the back storage room, pausing at the door. "And are you wearing new cologne or after shave today?"

Color blossomed across Henry's cheeks. "Well, yes."

Lois scrunched her face. "I'd stay with the old one." She proceeded on her mission.

Malcolm said, "Man, she's a handful."

Henry stole a fleeting look at the back room. "She gets things done you know; I think she's trying to make sure we stay afloat."

"Why Henry, I think you have something for strong women."

Lois returned and cogitated over the call list. "Let's send Malcolm to Nahum Y. Peachey's today, it will expand his understanding of the culture in the valley."

Henry responded, "I usually handle them Lois, you know with Zephaniah being a little bit…"

Her brown eyes narrowed. "Send him; it's time he meets that" —Lois hesitated before speaking deliberately— "family."

Malcolm asked, "Zepha' who?"

Savoring his bite of doughnut as would an epicure sipping vintage wine, Henry said, "Zephaniah Y. Peachey, Nahum Y.'s father and Happy John Y.'s uncle. When I first started here twenty years ago, it was Happy John Y.'s father, Sam Y. Peachey, who was the catalyst for the split of the yellow toppers from the black toppers."

"The great religious schism over buggy paint in Galloway County began then?"

Henry relished another morsel of doughnut. "No, but the

exodus caused a serious rift among the Amish in the valley. Sam Y.'s brother is Zephaniah Y. He refused to go along. Thought they were beset by the Devil over their differences."

"Differences over what?"

"Use of vaccinations, and prescription drugs, to name a couple. Zephaniah's older son, Moses Y., is the Amish alchemist for the valley."

"Well, maybe the next reformation will foster a little more creativity in picking middle initials."

"The two groups have shunned one another all these years."

"Even those in the same family?"

Henry swallowed the last of his doughnut. "Zephaniah Y. is the bishop of that particular group of old order black toppers. You might say he keeps a close watch over his flock, a *very* close watch."

"And they don't use vaccines of any kind?"

"Neither their family members or their animals. The prevention of disease is entirely in God's hands."

"Are they related to Phares Y., where I treated that tetanus horse last year?"

Henry reflected on the ceiling tiles for a moment, nodded. "Nahum Y.'s cousin."

"Medicine based on superstition, fabulous!"

"One of the women in that group died about five years ago from measles, she was eight months pregnant."

"Yet another religious cult suffering in the name of God."

Without shifting her focus from the paperwork, Lois' flat voice chorused, "With your overall appearance this morning, you'll be a magnet for the evil eye."

Malcolm turned her way. "What?"

Henry said, "Zephaniah Y. had a cow that should have had her horns removed at a much earlier age, lost an eye."

Malcolm got to his feet and drained the last of his coffee. Ready to escape into the open world, he opened the door and regarded Lois. "By the way Henry, I like the cologne. I imagine it arouses women who prefer musky scents."

"If you can't see fit to comb your hair, at least wear a hat," she retorted.

Malcolm presented two donuts in plain view in one hand while he pulled the door behind him with the other.

From Malcolm's perspective, the Amish were masters of subtle navigation. Points of reference, such as large boulders, neighbor's fences, and foundations of old barns that no longer existed suggested landmarks were devised for the pace of a horse-drawn buggy and not a motor vehicle. However, even at Lucille's habitual speed, Malcolm couldn't miss Nahum Y. Peachey's lane; the black billboard scribed in white text was an easy mark. The current sermon ranted, ***"And they became scattered, because there is no shepherd, and they became meat to all the beasts of the field, when they were scattered. Ezekiel 34:5"***

Lucille parked in the barn lot. He waved to who he assumed to be Nahum Y., seated on a hay mower, jolting behind his team on the far end of the field. After filling a pail of water in the milk house, Malcolm entered the barn and found three cows, two more than he had expected. The stench of the retained placenta was all that was needed to find his scheduled patient. The sulfurous odor draped the barn like an overcoat. A piece of paper hung on a baling twine over the middle of the aisle. Penciled letters scrawled, 'furst cow in barn needs cleened, other 2 cows need pregnacy chek'.

A low rumble echoed throughout the nearly empty stable. Malcolm spun about and confronted a massive head with deep-set eyes leering through the far side of four-inch iron pipe bars, solitary confinement for a psychopath. Malcolm scrutinized the latch to the pen gate; the adventure in John Y. Peachey's pasture was still fresh in his mind. He scrubbed the rear of the noisome cow with Betadine, squirted lube on his sleeve, and inserted his hand. He teased the placenta up the pelvic canal. With a sudden release, the slop slumped onto the ground, splashing his coveralls. Repressing the urge to gag, he kicked the rotted tissue into the barn gutter, peeled off the sleeve and fitted a replacement for the other cows, both of whom were pregnant. The bull had left his calling card.

Malcolm wrote Nahum Y. a note and placed the paper on an upside down plastic pail. The bull pawed muck onto his back. Tired of excessive bovine machismo, Malcolm said, "Give it up, I'm not bothering your girls." The brute responded by extending his neck and bellowing until the beams of the barn seemed to shake.

Straddling the gutter, Malcolm unzipped his coveralls and jeans and released the pressure on his bladder. Malcolm aimed his stream closer to the enraged beast. "How about that, bad ass?" The bull bawled and Malcolm roared back as he zipped himself up. "Stupid bastard," he muttered as he turned to leave. Facing him, not ten paces away, was a patriarch wearing a dusky coat and pants, leaning on a cane.

The deeply creased face tallied the countless days of exposure to the weather, the crooked spine, the ceaseless physical labor. The right eye was atrophied and opaque, but the pale blue orb on the left glowered at the farm visitor. Malcolm's face flushed to pale scarlet as neither spoke for an excruciating five seconds. Feeling as if caught by the principal while writing graffiti on a school locker, Malcolm endeavored to regain a professional demeanor. "The cows are all good, both pregnant and the other one cleaned." A gnarled hand rose with deliberation, aiming a crooked digit at the veterinary vandal, while the implacable eye remained focused on Malcolm.

Malcolm said, "Thank-you, have a nice day," and dodged away from the stable. Refusing to look back to the barn, Malcolm washed his boots in stony silence. He tip-toed into his seat, started the engine, and put Lucille into gear. Escaping the barn lot, the stooped figure in black slowly dwindled in Lucille's rear mirror. Malcolm grabbed the radio mic and announced to Lois that he left Nahum Y.'s farm, next stop Jonas P. Yoder's.

CHAPTER XIII

Malcolm continued to reflect on the one-eyed elder as he watched an antibiotic solution flow into the jugular of Jonas P.'s cow. He was alone with his meditations and mastitis patient; the other half-dozen cows in the herd were out on pasture. Sunlight filtered through the grimy windows of the barn, tinting the stable with a soft gray. Years of abandoned cobwebs draped from the white-washed ceiling beams, soiled with brown dust—typical white topper barn. A pair of border collies trotted into the stable, barked tentatively, and looked back to the entrance as their master entered. Stoic eyes in a weathered face peered from under the shade of a brown brimmed hat. *"Guten Morgen, Tierarzt."*

"Good morning to you Jonas, *Wie bist du?*"

"Good enough. It's a sunny day."

"Raking hay?"

"*Ja*, I'm giving the team a rest, thought I would come see here what is going on." The dogs laid quietly by the farmer's feet.

Malcolm noted the last of the solution trail down the hose from the bottle. "She seems to be better than yesterday. Is she eating?"

"*Ja*, she is."

Malcolm asked, "Do you have a new dog?"

Jonas P. waved his hand at the younger dog. "This is Fritz. Kip is getting older, it's time to start getting a new one to learn."

"Is Kip a good cattle dog?"

Jonas P. straightened his back. "The best." The older dog looked up and met the farmer's eyes.

Malcolm gathered his equipment to leave. "Hold the milk from that cow for four days, Jonas." As he stepped from the dusty stable into broad daylight, both of the collies ricocheted off his legs and galloped across the barn yard. Malcolm looked back at Jonas P. as loud cries disturbed the morning air.

"*Scheiße, die Schafe!*" Jonas P. exclaimed as both he and Malcolm jogged towards the pasture. Out of breath, Malcolm halted by the farmer's side and scanned the green expanse of the field. The border collies had wiggled through the fence and were streamlining towards a kinetic chaos of white fleece, black spindly legs, and anxious bleating. Jonas P.'s small flock of Suffolk ewes and lambs were desperately trying to escape a malicious force surging through the grass, a large black dog.

Malcolm flared into rage as the black beast flipped a young lamb into the air. Scrambling over the fence, he ran clumsily in his boots through thick clover. Jonas P. sprinted in the opposite direction, back to the barn.

The vile beast released the fallen lamb and targeted an older ewe, closing the distance on its prey in bounding strides. Just as vicious jaws reached out to snatch the sheep's hind quarter, Kip lunged into the side of the marauder. The smaller dog's momentum compensated for the disparity in weight and both dogs tumbled over the ground. Fritz refrained from closing into the fray, doing little more than barking. The black dog regained his feet and using his powerful forelegs, batted his smaller adversary to the ground. Kip struggled to fend off his aggressor, wiggling from side to side on his back, legs flailing.

Malcolm yelled at the top of his lungs as he desperately tried to close the distance. He spied a couple of smooth stones in the grass and scooped them into his hands. The first stone sailed an inch over the fiend's back, its fangs now latched onto Kip's shoulder. Malcolm steadied his balance, and with a windmill throw, whipped the lemon-sized stone through the air. The missile smacked the bony point of the hip of the larger dog, causing it to spring off its feet. Pressing his advantage, Malcolm pushed one foot over another and managed to free a boot. Launched in an eccentric arc, the black rubber was off aim but passed close over the head of the aggressor.

Waving his hands, Malcolm yelled, "Get out of here, you son of a bitch!" Fritz feigned another assault, driving the black dog to regroup 100 feet away. The beast growled while pale eyes sized up its human challenger.

A shrill whine pierced the air and a fountain of soil sprouted from the ground about the creature's hind legs. Spinning about, it sped towards the far pasture fence by the road, as another report from Jonas P.'s .22 cracked through the air. Malcolm's first impulse was to give chase. But winded and with one foot lacking a boot, he realized discretion was the better part of valor.

Kip rolled back onto his feet and limped towards Jonas P. The surviving sheep had swarmed into a small paddock in a far corner of the pasture. Shifting from one side of the enclosure to the other, wide-eyed ewes called repeatedly for their panicky lambs.

A pitiful moan arose a few feet away from where Malcolm stood. He tracked a mottled trail of fleece in the trampled forage. The lamb lay whimpering, a dull white blot in a bed of green. Malcolm sighed, "Dear God." The lamb's right ear was nearly severed. The face was ripped across the nostrils. The jugular was partially torn. Malcolm's head rang as he tried to register the animal's situation. Jonas P. came by his side, looked at the suffering animal and gazed out across the field. Straining to keep his voice, Malcolm closed his eyes. "I wish I could do something for this one Jonas. We need to end his misery. Would you like me to do it?"

Jonas P. shook his head. "No, veterinary, I will." Jonas P. raised the rifle, a sharp clap echoed across the pasture and the sobbing ended.

Malcolm swallowed with difficulty, struggling to let go of his anger. Jonas P. broke the silence. "I didn't think you could do much, veterinary. You shouldn't take these things so hard." He pointed to the flock up by the shed. "Maybe you can do something for them."

Contrary to Jonas P.'s request, Malcolm asked the question that controlled his thoughts. "Do you know who owns that dog, Jonas?"

Shoulder length hair under the brimmed hat shook imperceptibly. "Not really."

"Which way did he go out of the pasture?"

Jonas P. pointed towards the sun. "*Ostwärts.*" Jonas P. clicked the safety on his rifle and sauntered towards the flock. "I think there's a couple-three other ewes that might have been bitten. And maybe we should take a look at Kip." Malcolm gathered his jettisoned boot from the grass and followed the farmer.

The sheep in the paddock continued to circle in a huddled mass, snorting and stamping feet. Jonas P. hesitated from climbing the fence. "Don't know if I want to handle them much right now, being all upset. Maybe if they need treated, me and my boys will get them in the shed tonight."

Malcolm searched the flock for any sheep that needed critical attention, a difficult task as the bodies wove between one another. Rubbing his cheek, he said in a soft voice. "I'll leave some antibiotics for any with bite wounds, Jonas. I'll also leave some tetanus shots." Malcolm sighed, "I'm worried that dog may come back."

The farmer sized up the pasture. "I'll keep the flock in the paddock and shed at night." He laid his hand on Kip's head. "Kip got a little bit up as well, would you look at him maybe?"

"Of course."

Jonas P.'s wife and children had congregated in the barnyard. Rachel ran forward, asking her father, "Daddy, *was ist loss?*"

"Rachel, Go back!"

"*Bitte*, daddy."

Jonas P. relented. "Hold Kip for the *Tierarzt.*"

Rachel knelt down and cradled Kip in her arms. "*Gut hund,*" she whispered. She watched attentively as Malcolm's hands stroked the dog. Kip flinched as his shoulder joint was palpated.

Jonas P. laid his hand on his dog's head. "He wasn't near as big, but he is pretty slippy. Fritz, he's kind of young, he doesn't know."

Malcolm said, "That was a lot of dog to fight. Jonas, I want you to wash this wound out by the shoulder very well. Cut away the hair a little bit. I am going to leave you some antibiotic capsules you can mix in with some meat or something like that."

Jonas P. said, "Rachel will do that."

"Is he vaccinated for rabies?"

Jonas P. shook his head.

"I'm going to give him and Fritz a shot, just for good measure."

Malcolm found a bottle of red ampicillin capsules in Lucille and poured a handful into a plastic vial. He drew up a couple doses of rabies vaccine and grabbed the drugs for the sheep. Jonas P. grasped Kip, followed by Fritz, as Malcolm eased the needles under their skin. Malcolm handed the pills to Rachel. "One pill three times a day in some food, please." Tears welled in her eyes.

Malcolm studied father and daughter together with their loyal retainer. "Are you sure you don't know who owns that black dog?"

"Not really, no."

"If either the sheep or Kip don't seem to respond, call me. Sorry, Jonas."

"As long as illness stays in the barn and not in the home, *Tierarzt*." A panel of somber faces of all ages watched Malcolm from over the garden fence as he hastened to Lucille, banged the door shut and wheeled down the lane, unwilling to look at the pasture. The rest of the day, Malcolm robotically completed his professional tasks, but his mind was disengaged from veterinary practice.

At home that evening, Malcolm slumped on his couch. The remains of a grinder retrieved from his refrigerator had failed to stimulate his appetite. He tossed a meatball to Precious. How could Precious and Kip be from the same species as the black monster? That beast *couldn't* be someone's pet, more likely spawned from the deep hemlock stands in the mountains above the valley. Malcolm shoved his sluggish body off the couch, tossed the sub back into the refrigerator and grabbed a Yuengling.

CHAPTER XIV

Queen Anne's Lace and powder blue chicory bobbed in the sunshine along the roadsides. A cool breeze from the north spilled over the back mountain and pushed away the heat that stifled the valley earlier in the week. The fresh air energized Malcolm's spirits as he pushed Lucille up and over the front mountain and into the river valley.

Delmar Brubaker commenced milking at the scandalously late hour of seven o'clock, believing that cows gave more milk in the full light of morning. As he stepped out of Lucille, the buzzing from the vacuum pump informed Malcolm that Delmar was still at his task. He scuffed his boots from car door to back hatch. Rummaging for his stethoscope, he was nearly propelled through Lucille's back end by a blunt force on his back. Contorting his spine, Malcolm came face to face with an unruly mop of grey and white fur.

"Who in the hell are you?" Malcolm pushed on the chest of the English Sheepdog. The dog batted at him playfully and salivated on his hand. "Okay, that's enough!" Malcolm added a little more emphasis to his thrust. The dog rocketed into orbit around the Scout and upon returning, renewed his whimsical onslaught on Malcolm's leg. Malcolm raised a knee to deflect his assailant and grabbed his equipment between intermittent spars of jujitsu.

He escaped towards the stable while his new canine acquaintance cavorted about the barn lot in erratic arcs. The rhythmic whish-knock of the milking equipment as rubber liners opened and closed on cow teats brought Malcolm back to tranquility. A Haydn concerto creaked from an old cassette player because Delmar believed classical music 'relaxed the cows, so they would letdown their milk'.

Malcolm found Delmar squatting by a cow, washing her udder. "Morning Delmar."

Straightening up, Delmar pulled a red handkerchief from the back pocket of his bib coveralls and blew his nose. "Morning Doc. It's Missy over on the other side there, between Zelda and Lulu."

Delmar limped along with Malcolm. "If you ask me Doc, I think she's got too much fermentables in the ration. She was just fresh a few days ago." Delmar talked out of the side of his mouth as if keeping a phantom cigarette in place, complemented by a grey-blue striped cap reminiscent of Casey Jones. "I think she's a little ketotic, too. She needs more complexing carbozymes, so I drenched her this morning, get those good bugs in her intestinals immaculated, so I did."

"I guess that's possible", Malcolm said as he placed his hand on Missy's broad back. The canine nemesis returned to grasp Malcolm's leg, arch his back and gyrate his pelvis with gusto. Malcolm shook his leg free. "Get off me, you moron!"

"Pierre, get out here!" Delmar growled. The grey and white blur hightailed out of the stable, only to return and rip down the feed aisle in front of the cows.

"Who is that maniac, Delmar?"

The farmer shook his head. "My girl's dog. She returned home last month, but keeps him here rather than at her apartment."

The sheepdog was temporarily out of sight. Malcolm said, "Nice addition to the farm."

"Not a bad sort, dumb as bag of rocks. He gets a little antsy whenever anyone new arrives, then he settles down."

Malcolm returned his attention to his patient as the dog sped by the alley in the opposite direction.

Delmar tightened his round face. "You see, Doc, when a cow freshens, she gets into a negative energy field that consumes her beneficial antibodies. So I gave her a bottle of Maxi-Immune Gold Plus."

Malcolm sighed. Wonderful, another version of snake oil to

take farmer's money. "Is a whole bottle the usual dose, Delmar?" Malcolm asked while listening to the cows left side with his stethoscope, thumping her with his finger.

"Yup, for a cow at the start of lactation, like Missy here, a whole bottle."

Malcolm stroked the rear of the cow above the udder. "How much does that cost you?"

"Well, seeing how's I'm a local distributor, I get a special price of four dollars." In response to Malcolm's caress, Missy arched her back and dribbled some urine. Malcolm held a small strip in the urine flow and a patch on the strip turned a deep purple. He held the evidence up at eye level, "Ketosis, Delmar," and tossed the strip into the gutter behind the cow.

Delmar scratched his neck. "I told you, Doc, it's those fermentables."

"No doubt, but I better give her IV dextrose anyway." Delmar ambled up the aisle to revisit his milking. Malcolm restrained Missy's head with a halter and glimpsed who he guessed to be Delmar's daughter, resting a wash bucket next to Missy's neighbor, Lulu, to prep her for milking. She raised her head, revealing green eyes offset by auburn hair.

"Good morning, Doctor."

"Hello." His head buffeted into the milk pipeline, causing him to rub his brow. Malcolm recovered to get the bottle flowing into Missy's vein; his eyes wandered to the ceiling. The summer's second brood of barn swallows was overflowing their hovel of straw and mud attached to the beam. In two weeks, they would take wing and follow the sun to winter quarters.

The return of the auburn aura disrupted his musings. She carried a milking unit in her hand, the hoses wrapped about a tawny arm descending from a sleeveless grey tee shirt. The pastel hues streaming through the barn windows complemented her complexion. As she stretched to attach the hoses to the vacuum and milk lines overhead, her tee shirt lifted, exposing her planed abdomen. She bent under Lulu to attach the unit with fluid grace.

Malcolm absorbed the entire routine. She gave him a reproving look. "Is Missy going to survive?"

"Oh, yes, it's ketosis."

"Great." She held his gaze for a moment and went to fetch another milking unit. His eyes drifted along in the wake of her blue jeans.

Progressing down the line of cows, she said something to Delmar and dropped out of sight behind a cow. She reappeared with another milking cluster and headed towards Malcolm. He busied himself with a glance out the window. Sensing her sustained visual dissection, partly by peripheral vision, more so by telepathy, he turned his head.

Hand on hip, she appraised the bottle. "Is Missy finished, or does she need a second dose?" Malcolm glanced at the bottle—empty. He deftly removed the needle and rubbed the injection site with vigor. "It's a good thing you don't charge by the hour, Doctor." Her hand lifted the wash bucket. "I suppose Missy is well enough now after your expert care, that I can start milking her?"

Malcolm stumbled over the gutter to offer space for the milking equipment, doubling back to where Delmar was crouched by a cow.

"Delmar, you'll want to drench her with a pint of propylene glycol this morning and probably tomorrow morning as well." The farmer repositioned himself to face Malcolm. Malcolm perceived the strain in Delmar's face. It's always the knees that gave out first with dairy farmers, the price of milking twice a day, every day.

"I'll give her some more antibodies tonight as well, Doc."

"That's fine," Malcolm said indifferently as he scanned the stable.

"Hey Doc, while you're here, I've got some young heifers that weaned this week. I wonder if you've got time to look at them."

"Sure, Delmar."

"They're in the back shed."

Three young calves lounged in a pen, not more than eight weeks old. They nonchalantly rose to their feet as Malcolm intruded into their space. He hugged one about the neck and slipped on a halter.

The calf offered little resistance. Malcolm couldn't discern any abnormalities on his physical exam. While contemplating his next step, Pierre burst into the shed, followed by Delmar and his daughter. Delmar queried, "What's their problem, Doc?"

Malcolm traversed the spine of the calf with his hand. "I'm not sure, Delmar."

Delmar folded his arms. "I'm a little flummoxed myself. They were doing fine when we took them back here yesterday. I think their digestables have soured now that they're off milk." The farmer elaborated on his theory of antibodies and bacteria in the digestive tract of calves.

Malcolm stole a glance at the green eyes. They gazed back without wavering. Delmar was still chatting when one of the calves urinated. Malcolm heeded the brief distraction. The calf's urine was the color of red tea. He patted his pockets and found a syringe case. The warm flow bathed his hand as he caught a sample.

Malcolm wiped his hand on his coveralls and held the impromptu specimen to the light. "Delmar, have you given these calves any medications this past week?"

Delmar picked up his cap visor, scratched his head. "Why no Doc, can't say as I have."

"Have they been exposed to anything out of the ordinary?"

Delmar remained silent for a few moments. "The only new thing those calves got into was a good deal of water. They slugged down so much, they looked like a bunch of fat ticks. Drank it like there was no tomorrow."

Malcolm asked, "Why did they like it so? Is it cooler than the water you give them in the calf pens?"

"Oh, we don't much give them water in the pens, Doc. They get the milk and all."

"Delmar, given their size and the warm weather, they may have slugged enough water to cause some of their red blood cells to rupture."

"Huh, what do you know?"

"Hopefully, they've probably got past the engorging stage."

Delmar chuckled, "Kinda like kids eating so much candy they get themselves sick."

"Yeah, something like that. I'm going to collect a blood sample to check for anemia and spin this urine back in the lab. In the meantime, I think you should start offering water to your calves that haven't been weaned yet…in small amounts."

"Alright Doc, sounds good to me." Turning to his daughter, Delmar said, "Doc's a pretty darn observant fella isn't he, Jolene?"

"Yes pop, he's certainly aware of his surroundings." Her voice had a sarcastic trace.

Having rescued a portion of his self-esteem, Malcolm shuffled back to Lucille. Delmar and Jolene accompanied him.

Malcolm spied his bucket behind Lucille. "Where's my brush?" Casting his view to a wider field, his eyes rested on Pierre. The dog lay on his haunches, gnawing the bristles which rested between his front paws.

Delmar exclaimed, "That damn rascal!"

Jolene clapped her hands. "Pierre, bad boy!" She tromped over to Pierre, who relinquished his prize after a brief tug-of-war. She surrendered the brush to Malcolm. "Sorry about that."

Malcolm inspected the teeth marks, a fair part of the bristles were missing. He smiled, "No problem." He let his temporary leverage over the damaged brush set for a few seconds. "I'll go back to the shed and get a blood sample from one of the calves. I'll keep you posted on what we find, Delmar."

"See ya, Doc, thanks."

Malcolm called out, "Nice meeting *you* and Pierre, Jolene."

Her lips curved into an arc. "Keep on track, doctor," and she receded into the barn.

CHAPTER XV

Steel shod hooves hammered the pavement in double time and echoed through the old growth forest that bordered the edge of the road's shoulder. Despite the warm sun, the riders in the white-top buggy lowered their heads against the wind. Strong gusts swayed the trees with such animation that the limbs drooped towards the ground as if to sweep humans, horse, and buggy into the canopy.

The horse knew the way into town and was sensible about cars and other disturbances. Yet, the woman kept a firm hold on the reins; she found this stretch of road, devoid of habitation or farm field, to be perturbing. The wind distorted the shadows under the hemlocks, daubing the forest floor with only sparse bits of sunlight. The woman quelled her apprehension by thinking about their destination, the hardware store in Salty Dog. The garden was flush with tomatoes and beans and she needed canning supplies. Conversely, her teenage daughter slouched from the tedium of the trip and the company of her mother, scheming of a rendezvous with the neighboring farm boy. Perhaps she could make an arrangement after the ice cream social this Sunday evening.

A hefty cloud, goaded by the wind, obscured the sun and a wall of shade crept up the road and swallowed the travelers. The horse's ears snapped to attention as a pair of crows launched from a dead elm tree to the left of the buggy. Nostrils flaring, the Standardbred quickened his trot, while the woman scanned the colorless light under the trees. A presence, no more than a dark outline, kept pace with the buggy. She told herself it was a deer. The horse flung his head, and the woman needed all her strength to yank the reins.

"Behave!"

The horse escalated his struggle against the bit, swinging his rump to the side. The girl leaned on her mother and clutched her royal blue dress.

The matron's heart skipped a beat as a small branch clustered with leaves snapped off in a gust of wind and danced across the road twenty feet in front of them. She placed a hand on her daughter's knee and offered a thin smile. With a shrieking howl, a four-legged blur exploded from the underbrush and bounded towards them. Mother and daughter recoiled instinctively as the beast lunged at the side of the buggy.

Grasping the leather leads in one hand, the matriarch grabbed the whip next to the seat and lashed wildly across the face of the attacker. *"Gehst du, schmutziger Hund."* The crack of the whip jolted the horse into an uneven canter and the buggy lurched forward from one side of the road to the other. Fumbling to control the reins, the woman lost her hold on the whip and it tumbled out of her hands. The attacker gave chase, snapping jaws at horse's legs and wheels in alternating fits. The women rattled about in the seat like marionettes as the buggy jostled down the road with increasing speed.

Steering with one hand, Malcolm's forefinger from his free hand fluctuated between the rewind and fast-forward buttons on the cassette player. "Damn it, Lucille, don't you dare eat that Hendrix tape!" Precious loafed in the passenger seat and yipped intermittently whenever the Scout strayed across the center yellow line. "Cut it out with the backseat driving already, between you and this good for nothing stereo…" Leaning through a curve, Precious' safety alarm amplified to repetitive barking. Malcolm peered over the dashboard. "What in the hell?"

Adrenalin galvanized his senses as he beheld a white top buggy bearing towards him, reeling behind a galloping horse. Malcolm grabbed the wheel with both hands and narrowed his eyes. "I can't believe this!" The buggy had an escort—a large black dog. Malcolm flashed with anger, not any black dog, *the* black dog.

The Standardbred pressed against the lead lines and harness, wanting desperately to break into an unbridled run. The horse veered to the left, directly ahead of Lucille. Malcolm drove his right foot into the brake pedal as the horse bucked and twisted to the right. Precious rolled forward, her front paws braced on the floor. Tires screeched to a stop. The buggy wheel hub grazed the front quarter panel of the Scout with a repulsive metallic squeal. White spokes streaked past the driver's window a foot and a half away from Malcolm's face.

Malcolm trundled out of Lucille to gain a better view of the trajectory of the panic-stricken horse. The weight of the carriage shifted towards the outer edge of the curve as rapid-fire hooves drummed the asphalt, but all four wheels managed to maintain contact with the ground. Malcolm turned to find Precious with bared incisors and rigid hair, transfixed by yellow embers in a coal-black face in front of Lucille.

Malcolm reached onto the wheel and gave an abrupt blast from the horn that startled the black fiend, while he wasted no time to shimmy back into his seat. He simultaneously shoved the gear shift and punched the gas pedal. "You need an attitude adjustment, my friend." Precious vocalized her approval, paws on dashboard. His adversary jumped to avoid the front bumper and scrambled off the road and across a ditch.

Malcolm halted and sized up the dog through his windshield. The beast bayed unremorsefully as it vanished into the forest. Searching the rear-view mirror, Malcolm mumbled, "Better check on the women in the buggy." Lucille spun about in a U-turn.

Malcolm sped up the road and around the curve to find the buggy still barreling down the road. Malcolm thought, got to get ahead of that horse. He accelerated, realizing that a tight curve, dropping down a grade, was less than a mile ahead. Catching up to the rear of the buggy, he looked for an opportunity to pass, but was denied twice as the horse swerved across the center line.

Finally, as the horse swung more towards the shoulder, Malcolm saw his chance and cut into the left lane. A white pickup popped out from around the curve ahead and closed the distance to Malcolm with alarming speed. With one last push on the pedal, Lucille banked back into the right lane in front of the buggy, as the horn of the pickup blared angrily. Malcolm continued to distance himself from the buggy and just before the curve, jammed on his brakes, placing Lucille horizontally across the right lane.

Getting out of his seat, Malcolm said to Precious, "Stay low, this horse has had enough of dogs today." He calmly stood by his front bumper, arms extended from his sides, but not moving. Bracing himself, the heavy wheels closed in. The Standardbred slowed to a trot and reared onto hind legs, ten feet from Malcolm. Malcolm talked quietly to the frightened animal and after a few moments approached with care. The horse was covered in sweat, whites of the eyes visible. Malcolm continued to soothe the horse and gently reached out to take the reins. He stroked the muzzle and glanced at the riders.

The faces of both women were drained of blood and their dark kerchiefs hung loosely about their necks, still attached by the knots. The girl shivered, clutching her mother's fabric.

Malcolm asked, "Are you okay?"

The matron replied, "*Ja*. That evil dog."

"Yeah, evil dog indeed."

An oncoming car slowed down as it drew even with the horse. A driver in a baseball cap called out, "Is something the matter here?"

Malcolm said, "Better than it was. If you don't mind, would you pull off the road and hold onto this horse while I give him a sedative?" Malcolm turned to the women, still frozen in place on the buggy seat. "I'm going to give him something to calm him down. He can't go any faster than a walk the rest of the trip. I'll drive slowly in front of you to help pace him into town."

As Malcolm returned to Lucille to gather his injection, his mind was made up. Time has come to find out who owns that dog.

CHAPTER XVI

Malcolm's shoes squeaked on the glossy pavement as he descended into town. Uneven clouds dabbing the ridge tops were preparing to wring out another rain shower. He entered the town square and waved at a couple of village patriarchs, planted in lawn chairs inside the open garage door of the firehouse, while punch-drunk moths spiraled in the light above their grey heads. Malcolm bore left and walked another half block, past the hardware store and the bank. Coming upon the Dead Bear, faint tendrils of volatile fumes twitched his nose. Someone was working late at the auto body shop next door.

Malcolm located Shannon and his companions in the waxy glow of the bar. Shannon cried out, "Doc, hey, what do you say, man? Come the hell on over and have a beer with us." Malcolm stopped by the bar, ordered a bottle of Yuengling and four more for Shannon and his comrades.

Shannon smiled. "We appreciate that, Doc. Don't we Jimmie? Doc, this here's Ronnie and Dozer." Malcolm recognized Ronnie as Shannon's dog breeding antagonist; his silver hair suggested that he was a good deal older than the other three. His choice of apparel, bowling shirt and slacks, was common to both the present gathering and the night of the dart match. Dozer was easily six and a half feet, bearded, and probably could have tipped a scale at three hundred pounds. Malcolm asked, "What's going on?"

"Just hanging around with the boys, although we're in for a little pinochle later," Shannon replied. "How about you, Doc, play pinochle?"

"Spent more than enough time playing in college."

"You the veterinary ain't you?" inquired Ronnie in a gravelly voice.

"Yes sir, I am."

Ronnie leaned forward, scotch tape held one of the temples of his eyeglasses together. "I always wanted to know if pigs are smart."

"Well, they ain't smarter than a dog, anyway", Shannon said.

"Some people say they are," Ronnie retorted, lighting a cigarette.

Malcolm said, "Animals can be smart in different ways." He hesitated. "Horses learn their way home from one place or another. Dogs can be trained to be Seeing Eye Dogs."

"And ride bicycles in the circus," Shannon added.

"That's goddam chimps that ride them bicycles, dogs jump through hoops," argued Ronnie.

"Chimps my ass, I've seen them on TV", defended Shannon, "and besides those are hoops of fire, man."

Malcolm plunged in. "Don't know much about *that*, gentlemen. But I can tell you something about pigs." The debate ceased. "Where I was a student, there were lots of pig barns at the university. Because of the cold winter weather the barns were heated. Of course, this cost lots of money to keep warm air blowing on the pigs all night. So, the barn crew set up switches for the heaters in places that the pigs could reach."

Ronnie nodded and looked pointedly at Shannon. Malcolm continued, "The pigs would turn on the heaters if the barn got cold and after a spell, a timer would turn the heat off. When the pigs got cold again, one of them would turn the heat back on, and so it would go." By the look on their faces, Malcolm might just as easily have transformed his beer into a vase of flowers.

"What did I tell you?" Ronnie swept Shannon's shoulder with his fingertips. "No damn dog is gonna walk over and turn on the thermostat in *your* house."

Shannon said, "They could do that. You've seen that TV commercial with the dog getting the beer out of the refrigerator."

"That's a bunch of fake bullshit on TV."

"Fake my ass. If a dog can get a newspaper, he can fetch a beer. I could train Buster to do that."

"That hound of yours don't even come when you call him."

"Speaking of fetching, I'll go get another round," Malcolm volunteered.

He was greeted again by the bartender. "Shannon and the boys educating you?"

Malcolm looked over his shoulder at the lively discussion. "They do seem to have rather strong beliefs, even if not much based in fact."

She handed him fresh bottles, stretching a tattoo of a red rose along her forearm. "I'm Katy."

"I'm Malcolm. Nice to meet you."

Her blue eyes flicked across the room and returned to Malcolm's. "If ever you see fit to avoid the sages during one of your visits, come have a seat sometime."

Malcolm grasped the beers. "Probably would be more productive." He shuffled back to his cohorts, glancing back at the bar.

Ronnie had his arms raised as if offering a benediction. "Jesus cast demons of madness into a herd of pigs, handpicked for their Holy mission."

Shannon said, "And they went and ran themselves off a cliff, didn't they?"

Malcolm pictured evil spirits bursting from the mind of a lunatic and infesting others like a contagion. An uncomfortable thought crossed his mind. Not unlike the rabies virus. He turned to Shannon. "I wonder if you guys can help me with something?"

"Why sure, Doc."

Malcolm waited for their attention to settle in. "Jonas P. Yoder's sheep were attacked. A lamb was killed."

"Do you mean Jonas P. the white topper up on Shaneytown Road?"

"That's the one."

"Sweet Jesus wept."

"A large, black German Shepherd-type did the killing. Any of you guys know who might have a dog like that up in that part of the valley?"

"You seen the dog yourself, Doc?"

"With my own eyes."

Shannon looked at Jimmie, then at Ronnie. Dozer spoke up. "John MacDougall has got a dog like that. I deliver fuel oil to his place. I won't get out of the truck if that dog is loose. MacDougall can hardly control the damn thing himself."

"How close does Mr. MacDougall live to Jonas P.'s place?"

Shannon said, "A little less than a half mile up towards the gap."

Malcolm sorted out the circumstantial evidence. "I know Jonas P. won't make a fuss over this. But I would like to find out if this is the same dog."

"I'd be mighty careful of that plan," Dozer warned.

Shannon rubbed the back of his neck. "Doc, MacDougall ain't right if you know what I mean. The whole family is rather, uh"— Shannon rotated his head to get a full panorama of his surroundings—"colorful."

"Colorful?" Ronnie's voice pealed across the clatter of the bar. "Colorful? That clan is a bunch of damn maniacs."

"MacDougall is a bit of a pisspot, Doc, I'd stay away," Shannon echoed.

"I'd sure would like to check that dog."

Ronnie leaned closer to Malcolm. "The place is haunted."

Shannon closed his eyes and made a face. "Oh, for God's sake."

"Go ahead, tell him...."

Shannon spoke in a barely audible tone. "That upper end of the valley is a little less, uh, civilized. You know how the ridges come together there, kind of poking their fingers into the valley?"

"Yeah," Malcolm responded.

Shannon held out his left hand and separated his fingers for effect. He squeezed his ring finger with his other hand. "Especially this second spur from the south, Black Log Hollow. People just don't go up there much, even to hunt. Peculiar mood to the place

Doc, kind of like an Indian burial ground or something. Strange things happen from time to time."

"Like what?"

"Lights, noises, missing folk and livestock."

"Chopped and hacked," Ronnie said nonchalantly. Malcolm stopped sipping his beer. "It's the God's truth," Ronnie continued. "It all started with Eliza Mae MacDougall. She went crazy while her husband was away fightin' in Grant's army. Came back after the war and one night she hacked him into pieces with an axe. Poor bastard survives the Rebs, all to be butchered at home in bed."

Malcolm squeezed his bottle. "Good God."

Shannon said, "Family was moonshining for many a year. Hell, rather profitably at that. Ol' Jebediah ran a good business, kept it respectful of the law, no trouble, paid the right people."

"Back in Cook County, Illinois we called that bribery."

Shannon flashed crocodile teeth. "You got it, Doc. Jeb had a pistol of a son, Trotter. He was a wild thing, the usual misbehaving stuff. Then out of the blue, one day he goes crazy like his great-grandma. Some say that Trotter's and Eliza Mae's ghosts wander about as unsettled souls. They were buried in the family plot on the homestead instead of in a proper God-fearing cemetery."

"Don't forget Archie, younger brother of John MacDougall," Ronnie added. "Came back from Vietnam, crazy as a bat. People would see him stalking outside their homes at night, peeking in their windows."

Shannon dismissed him. "People say lots of shit."

"Yeah, well those animal rituals stopped after he vanished."

"Animal rituals?" Malcolm asked.

"Folk's pets gone missing, find them all gutted and dressed like a deer, hanging in trees, sacrifices for witchcraft he brought back from Asia," Ronnie said under his breath.

Shannon said, "Mind you, that was some years ago, when we were all younger colts."

"Where did Archie go?"

Shannon shrugged. "No one knows. Some say he got in trouble

with the law out west somewhere."

"I think MacDougall keeps him hidden in his basement," Ronnie said.

Shannon tapped Ronnie on his chest. "Maybe MacDougall lets Archie out at night like a bat. So he can come over and peek into your bedroom window. Seeing your skinny bare ass would be enough to scare him and all the other ghosts back to the basement for at least a year."

Dozer, Shannon, and Jimmie laughed out loud. Malcolm, on the other hand, remained pensive. "This makes sense with what Dale and Sally Calhoun told me some time back."

"About what, Doc?"

"I was at their farm, happened to see MacDougall's white truck go by, Dale went on a rip about him."

Shannon's canny eyes stayed on Malcolm while he took a swig from his bottle. "Bad blood between those two clans…runs back aways. I wouldn't get in the middle of that, you bein' new around here and all."

Thinking it was time to go home, Malcolm wished them good luck with the Pinochle game. He threaded his way through the still crowded tables and returned to the bar to drop off an empty bottle.

Katy was filling a glass at the tap. "Leaving already?"

"It's been a long day. I'll have to take you up on conversation some other time."

"I'm sure we could find lots of things to talk about."

"Good night." Malcolm drifted to the exit and eased around a couple embracing inside the doorway. Ronnie's voice carried across the bar. "No pig would ever attack sheep like a damn dog!"

Outside, Malcolm lifted smoke free air into his lungs. A hand gently clasped his shoulder. "You be careful of getting mixed up in all of this, Doc. The MacDougalls are a bad bunch and they got plenty of kin about, too."

"Thanks, Shannon. Where's his place, anyway?"

"On the Shaneytown Road, to the right at the "Y" past the Calhoun place, just like you go to Jonas P. Yoder's. He'd be the third place on the right, second place after Jonas P.'s, off the road a piece. They've got a big old rambling lodge of a place. Their family had money once."

"Because Seagrams bought out their still?"

Shannon looked at him with a brief moment of sobriety. "You watch yourself. And don't go there near nightfall."

The soles of Malcolm's shoes slapped a dull percussion on the roadside. He stewed over the reputation of clan MacDougall and their proximity to Jonas P.'s flock. A soft breeze accompanied him home with the scent of rain and a sense of isolation.

CHAPTER XVII

Drizzle seeped from a leaden shroud that hid the heads and shoulders of the ridges, amplifying Malcolm's misgivings. *Should I have told Henry about the runaway buggy and now this trip?* Lucille's wiper blades smeared a mixture of dust, water, and insects across the windshield. For once, he wished the drive to his destination would take a little longer. He passed Dale's farm before the "Y" then bore to the right. A specter of shredded white fleece hovered in the clammy air over Jonas P.'s pasture. Malcolm pushed the accelerator harder. He passed a small farmstead on the opposite side of the road; a place he couldn't recall ever visiting. Further on, a dirt track branched off from the right side of the road and disappeared over a rise and into a screen of trees.

Malcolm halted by the entrance, leaned across the seat and rolled down the passenger window. A neglected mailbox listed away from the road, identifying the habitants with 'MacDougall' coarsely printed in black letters. Rain-laden brush and saplings drooped over the lane, narrowing the passage into a tunnel, pock-marked with mud puddles. With a deep breath, Malcolm turned the wheel.

Even at a crawl, Lucille jerked up and down with each dip and bump. As he rose over the crest of the lane, two spurs of the ridge came into view, curling in an arc around the entire back half of a large clearing. An imposing three-story structure, fortified with walls of massive brown logs and grey mortar, dominated the space. Stone chimneys towered over the roof at either end of the home, and a covered porch, supported by field stone columns, extended along the entire front. An unkempt barn and shed stood further beyond, the boards half-covered by ivy. In a back lot, behind the barn, a couple of rough-looking Black Angus steers ranged about a brushy pasture. To Malcolm, the ridge spurs appeared as brawny arms, gathering the entire homestead into a subterranean lair.

An old Ford coupe, coated with more rust than faded black paint, lay broken in a bed of lanky weeds. The relic was centered where the lane fanned out into the clearing, so that a visitor was obligated to steer either towards the barn on the right or to the house on the left. Malcolm edged by the car, glancing over to see a cracked windshield and what seemed to be scores of bullet holes throughout the hood and side panels. A familiar white Chevy pickup sat in front of the house. More critical for Malcolm, the black dog was tied under a shade tree beyond the porch steps, clearly making the caller's presence unwelcome. *So, this is your home after all, you bastard.*

Lucille coasted to a stop in front of the porch, a corroded washing machine rested at an angle, bending the lone shrub in front of the porch—heaved and left in place like a dystopian lawn ornament. Mentally rehearsing his introduction one more time, Malcolm opened the door. The radius of packed dirt around the tree bore evidence that the full length of the dog's fifteen-foot chain was tested regularly. Vicious jaws snapped uncomfortably close as Malcolm climbed the porch steps. The sterile backdrop of the porch without a single piece of furniture validated the lack of hospitality.

He was surprised how timidly he knocked on the door. The portal swung back and he was face to face with a man of about fifty-five, six feet in height, with a heavy abdomen and a bull neck accentuated by a short buzz cut of hair. Blue eyes radiated enmity behind horn rimmed glasses.

"Mr. MacDougall?" His host offered no response. "I'm Malcolm Cromarty, a veterinarian here in the valley, been here a couple of years." Without acknowledgement, McDougall pulled a pack of Marlboros from his shirt pocket, pushed one in his mouth and clicked a lighter. Malcolm persisted. "I do a lot of farm work in the area, and I was at Jonas P. Yoder's a few days ago." Malcolm raised his hand to point towards the tree line beyond the barn. "You may know him? He lives just west of you a little bit."

MacDougall remained passive, the dog possessed. Malcolm's uneasiness increased in proportion to his heart rate. "Mr. Yoder's sheep were attacked by a big black dog that is similar to yours in appearance. I thought I might stop by…"

"So that dutchman says that my dog attacked his sheep, did he now?" MacDougall said as alcohol flowed from his breath.

"No sir, he didn't, I was only trying to find dogs in the vicinity that match the description of the one that attacked the sheep and your dog is one possibility."

"Who told you it was *my* dog?"

Malcolm raised his hands. "No one. I was surveying some of the folk who have dogs around here and your name was given. Your place is close to Jonas P.'s."

"How did you know about my dog? Did that asshole busybody Dale Calhoun tell you?" Stretching his neck, MacDougall looked past Malcolm towards Lucille. "You do his veterinary work? That crappy paint job on your Scout looks familiar."

"I do his herd work, yes, but I didn't get any information from him. I'm actually going to go stop by a couple of other places," Malcolm lied.

MacDougall took a long drag on his cigarette; the glow was the sole point of warmth in the sultry mist. The dog continued to propel against the limits of the chain and his master made no effort to cease its obsession. Smoke drifted from MacDougall's nostrils and into Malcolm's face. "Well, if that goddam dutchman has a problem, he can come see me."

Malcolm played his trump card. "Mr. MacDougall, I would like to offer rabies vaccinations to any of the potential suspect dogs and check them in about ten days, just to make sure they're alright. I'll give the shot for free, no expense to you. Then your dog would be protected. We got notice from the state that rabies is a growing problem, especially with raccoons and there was a rabid cow in the valley not too long ago."

MacDougall lowered his head, pouted his lips, lifted his face to stare at Malcolm and took another long drag on his cigarette. "I don't think so."

"But Mr. MacDougall, your dog is at risk of getting rabies. It's a terrible disease. Vaccination will protect you, your family, and your dog."

"No other family here. Is the shot required by the law?"

"No, it's a good insurance policy."

"Then I guess I'll decline."

"But you need to get your dog vaccinated."

MacDougall's eyelids narrowed. "The thing of it is, I don't need to do a goddam thing. My dog is none of your fucking business."

Malcolm said, "I was just looking out for everyone's best interest."

"Whose best interest, Mr. Cromarty? Youens have no right saying my dog attacked that goddam dutchman's sheep. I say he didn't. I would suggest, *for your best interest*, that you mind coming here again. You want to give my dog a shot? Go give him one yourself. Come here again and I might be tempted to let him greet you off his chain."

But for his preoccupation with the canine malice, Malcolm might have tripped backwards down the steps in response to MacDougall's verbal assault. He grasped Lucille's door handle and cast a quick look back at his host. Cigarette in mouth, he was tussling his dog as it stood on hind legs. MacDougall shouted, "By the way Mr. Cromarty, tell Dale Calhoun to go fuck himself."

Malcolm reversed Lucille, being careful to avoid the white truck. As he turned to escape down the lane, MacDougall's voice rang out, "Hey Mr. Busy Body Vet, how fast can that piece of shit Scout go?"

The unbound dog took its cue and leapt after the vehicle, snarling and running alongside. Malcolm jolted over the ruts with considerably more vigor than during his arrival, sorely tempted to play bumper cars with his pursuer, who badgered him the entire length of the lane.

Frustrated, Malcolm wheeled Lucille onto the road, watching the demon recede in the mirror as he gunned the accelerator. A wry smile creased the corners of his mouth. He was disinclined to relay MacDougall's message to Dale. He doubted it would offer Dale any new sentiment on behalf of his neighbor that Dale hadn't already heard.

CHAPTER XVIII

The previous day's mist thinned into a veil of hazy sunshine as Malcolm made the rounds in the lower end of the valley. Needing to cross through Salty Dog between two farm calls, he drove into the village center and parked in front of the township hall.

The Kilderry Township Police Department, in total, consisted of Chief Cal Geisfelt, his nephew Tommy, and second cousin, Frank Miller. Most of their career was dedicated to keeping drunk drivers off the road, correcting juvenile delinquency, supplementing the township treasury with speed traps on the U.S. highway that skirted past town, and defusing domestic disputes. The screen door creaked to announce Malcolm's entry. Chief Cal was stretched back in a wooden office chair, feet on the desk, and elbows on armrests. His chiseled face and wiry salt and pepper hair were true to form with his service as a sergeant in Korea. Beyond the desk, two vacant jail cells, uniformly colored in institutional sea green, awaited the customary weekend arrival of alcohol-inspired miscreants.

Cal lowered his feet and lay down his fishing magazine. "Afternoon, Doc, how are things with you?"

"Fine thank-you, Chief, and you?"

"Just thinking about catching the big one. May go up to the dam this weekend, try for some small mouth."

"Good sized ones there?"

"The DNR has been stocking the lake for years. Do you fish, Doc?"

"A little fly fishing now and then, but I catch myself up more in the line than I do trout."

"Well, Tommy is pretty fair with tying his own flies, uses special feathers. Has a pretty good eye for picking the right pool in a stream, too."

"I imagine he doesn't try around here on opening day in the spring, my impression is that more fish are trampled than caught."

Cal chuckled, "You got that right, Doc." There was a lull, Cal's eyes settled on Malcolm.

"Chief, did you hear of the dog attack at the Jonas P. Yoder farm a few days ago?" Malcolm figured that most of the valley knew this news by now, but he asked anyway.

Cal sat back in his chair. "Yes, I surely did. That was a terrible thing. How many sheep did he lose?"

"One that I know of."

Cal pressed his lips together, shook his head and hummed. "Did Jonas P. get hurt?"

"No, he's alright. But his dog got a few lacerations trying to defend the flock."

Cal placed his hands together at the fingertips, keeping his elbows on the armrests. "So, it seems you witnessed the attack?"

"Sure did. And considering the attitude of the dog and owner, I would say Jonas P. has a problem with his neighbor."

Cal sat up in the chair and rubbed his cheek with the palm of his hand. The chair needed some grease. "What do you mean by the attitude of the owner?"

"I went up to John MacDougall's yesterday, to see if it was the same dog that I saw in Jonas P.'s pasture." Cal opened his eyes wider and cocked his head. Malcolm continued, "I had to be sure it was MacDougall's dog."

"You went up to the MacDougall place?"

"Yes."

"You're absolutely sure it was his dog?"

Malcolm chose not to reveal the buggy incident to Cal. "Absolutely." Cal looked out the window, nodded, and redirected his gaze on Malcolm. "Yesterday, I told MacDougall I was making the rounds to vaccinate dogs in the area for rabies, kind of an excuse to visit his place. You know we're having problems with an outbreak of rabid raccoons in the area."

"So I've heard."

"Well, I didn't think much of the possible connection when I got a close look at the dog yesterday, but after thinking about it last night, it occurred to me that this dog is unusually aggressive."

"Which means what?"

"What if this dog has rabies? It isn't vaccinated. I should visit the dog again in about ten days to make sure he was normal."

"Ten days?"

"If the dog has rabies, it would be shedding virus in saliva only in the terminal stages of the disease, perhaps two or three days. So if a dog bites someone you observe the animal for ten days, to be on the safe side. Of course, if abnormal behavior develops, the dog has to be put down and tested."

"But this dog didn't bite Mr. Yoder?"

"No. But Jonas P.'s collie was attacked as well as some of the sheep. Jonas P. hasn't been vaccinating his dogs for rabies, either."

Cal shook his head.

Malcolm said, "Can't you go up there and get the dog? Find a place for it somewhere, like the county animal shelter?"

Cal exhaled a long, exaggerated breath. "Don't know if I can do that, Doc. The thing of it is there was no human contact and Jonas P. didn't file a complaint."

"You know being Amish, he wouldn't anyway."

"I suspect, as do you, MacDougall's dog is a damn nuisance, but I would need a court order. And I couldn't get one without a complaint."

"Maybe you or Tommy could go by in a couple of days, just to see the dog?"

"If MacDougall gives us permission, but I can't demand that."

"This has got to be done."

"Now Doc, you let us take care of this problem."

"I hope you do."

Cal's mien transformed from congenial to icy. "I know you work with the Calhoun boys on their farm. I know too well what they think of MacDougall. I'm warning you; don't take sides in that feud."

Malcolm grimaced, thanked the chief for his time and made for the door.

"Doc!" Malcolm stopped and rotated to face the chief. "Most of the Amish are good folk. But they're set in their ways." Cal paused. "I wouldn't pick fights with the bullies of the valley on their behalf."

"Thanks, Chief." Biting his lip, Malcolm sought refuge in Lucille. "What's with people around here?" he muttered as he started the engine.

*****

Later that week, he was venting his aggravation with surly dog owners while sitting in Bobby's couch. Bobby held out a second Rolling Rock in his hand. "You need to party a little bit, might lower your blood pressure."

"Not another night of dart tossing?"

"No, man, a real party. Remember last year when I tried to get you to go with us to that big barn dance? Blue grass band, pig roast, keg?"

Malcolm examined the bottle in his hand. "Yeah, that was about two months after Carrie left me, wasn't much in the mood."

"Well, it's coming up again, kind of an annual event."

"You know these people well enough for me to drop in?"

"Marcy went to school with one of them. Bring along something to eat or drink for the kitty."

"When is this affair taking place?"

"Next Saturday." Bobby countered Malcolm's scowl. "Well I was going to call you, just didn't get around to it yet."

"I don't know about this."

"Look man, you meet Marcy and me here and we'll get you through the door."

Malcolm took a swig of beer. "What time?"

"Oh I don't know, once I get cleaned up after chores."

"Okay."

"That a boy," crossed a lazy fist at Malcolm, "Kicking and screaming we'll get you out to have a good time."

"By the way, you wouldn't know why the barmaid at the Dead Bear personally introduced herself to me the other night?"

"Me? No, why?"

"You're a bad liar, don't try politics." On his way home, Malcolm fretted over his new social commitment. The last time he had square danced was in junior high—an awkward week of commingling and self-awareness. Memories such as that reminded him why his adolescent years were a deplorable experience. At least he now had something else to occupy his mind besides MacDougall and his vicious dog.

CHAPTER XIX

Blueberries melded with a flaky crust in Malcolm's mouth, the remnants of a pie bought at Jonas P. Yoder's. Was it Jonas P.'s wife, or perhaps his daughter Rachel, who crafted this? Did young Amish girls like Rachel ever dream of new horizons, beyond the mountains that hemmed in the valley? How many of her white top Amish peers ever migrated, let alone traveled, away from here?

Out his kitchen window, Malcolm appraised the brooding ridges across the valley. The titans were more than a physical boundary of rock and shale, they were a psychic barrier as well. Malcolm tapped his fork on the pie plate. People were born in the valley, raised in the valley, educated by those who were native to the valley, worked, procreated, celebrated the joys and suffered the miseries of life in the valley, and ultimately died in the valley. A thought flickered in his mind. Will I be buried here?

Crickets serenaded the late afternoon sun as it exited behind the ridges. Debating whether to have another piece of pie, Malcolm tossed the crust to Precious. The phone rang, Malcolm stared at it with little enthusiasm. On the third ring, he sighed and put the receiver to his ear.

"Doc, this is Dale. The boys in the fire hall been saying MacDougall's dog needs looked at."

Malcolm ran his hand through his hair. Forget the Dead Bear, next time I need discrete information, I'll advertise on the local radio station. "Dale, I believe the dog is mean spirited, but I'm afraid that's his everyday behavior."

"That hound and his pot licker master are a threat to good, decent folk and with it having rabies and all…"

Malcolm's nearly choked. "Hold on Dale! Let's not go spreading wild rumors. Chief Cal told me not to…"

"Cal is such a bonehead. He don't want to get involved 'cuz he wants to be fishing or sit in his office and think of fishing."

"Dale, I had my chance to see the dog and Mr. MacDougall was adamant about no repeat visits."

There was a pause on the other end. "You want to see him again don't you?"

"Dale, aggressive dogs attack sheep all the time."

"This ain't about an aggressive dog, Doc. This is about MacDougall having his way with them dutchmen. Everyone knows he hates them."

Malcolm grabbed a tumbler from his cupboard and stretched for the ice tray in his freezer, time for the single malt.

Dale continued, "Me and Jack can help you see the dog again."

Malcolm clutched the ice tray, oblivious of the cold burn on his hand. "What? I'm under the impression he doesn't care much for you either."

"Not us Doc, you. We'll help *you* get another visit."

A neuron of negativity fired in Malcolm's mind. Not comprehending why, he asked, "How would this happen?"

"Easy, Doc. MacDougall goes into town most every evening to get himself smokes and groceries. Jack will watch for him at the end of my lane, he'll call me on the radio and I'll relay the call on the phone from my home. You'll have yourself time to get up to his place, look at the dog, and get out. You'll be long gone by the time MacDougall gets back."

An entire bank of 'no' neurons overloaded Malcolm's cerebral switchboard. "I don't think it's a good idea, Dale."

"Doc, as a vet you know that the whole damn valley is fretting over this rabies thing, ever since that dutchman's cow died."

Malcolm remained silent, paced on the linoleum and twirled the phone cord in his fingers.

"We're just fixin' to help you do the right thing."

Uncertainty spread in Malcolm's mind. What was the right thing to do in this valley? "Tell me how this works again?"

For the past twenty-four hours, a cold hand had clutched Malcolm's abdomen, twisting his entrails into knots. Precious barricaded the door. "Woof!"

"No, you can't go on this trip."

"Woof!"

"Precious, I said, no!" She sat on her haunches, refusing to surrender her position. Malcolm scratched her between the ears. "I don't want you to get in the middle of this." He slid out the door, blocking her exit with his legs. He pressed himself into Lucille and turned the ignition. The engine uttered an unproductive coughing spasm. His foot slammed the gas pedal while he twisted the key repeatedly. "Come on!" The Scout sputtered to life. Precious appeared in the window, baying through the screen.

Trying to ignore her, a sense of betrayal came over Malcolm as he guided Lucille down the driveway and wasted no time rolling through the town square, turning east towards the upper end of the valley. His only encounter on the road to Dale's farm was an Amish buggy, pulled at a slow pace by a horse with a droopy head. Jack Calhoun was perched in his truck at the end of his brother's farm lane, his face molded into a cryptic leer while Dale fidgeted in the passenger seat.

Dale called out, "MacDougall should be in town by now. Jack picked me up so we can go keep an eye on him together. We figure he's at the corner market by the firehouse."

"I'll only need five minutes, Dale."

"Well, just in case, we'll kinda get in his way if he comes back too soon."

Malcolm's eyes rested on Jack's left arm on the edge of the truck window. A crude tattoo of a skull with knives as crossbones decorated his bicep. The gun rack behind them had a shotgun.

"Don't worry, Doc, we'll say hello. He'll always make time to say hello back." Jack's mouth opened into an unpleasant smile.

Malcolm's lips tightened. "Alright, I'm going."

Jack extended his arm out towards Malcolm. "Touch the rabbit's foot for good luck."

Malcolm inspected the small piece of brown fur, thinking it would have looked better on the rabbit. He reached his hand towards the talisman, just shy of contact.

Jack wiggled the chain and the suspended foot oscillated. "Go on, touch it!"

Malcolm obliged.

Jack shifted his truck into gear as Dale yelled, "If you ever want one of your own Doc, Jack will make you one."

Jack's eyes gleamed. "Better to be lucky than good and better still to be bad and lucky!" Spraying a shower of stone behind the rear wheels, the Calhoun brothers rocketed down the road to town.

With far less bravado, Malcolm headed in the opposite direction and turned right at the "Y" in the road. The plan seemed simple enough. Malcolm would capitalize on MacDougall's temporary absence and observe the dog; a remote exam from the car seat to insure normal behavior, and leave. A disturbing thought crossed his mind. He minded MacDougall saying that no other family lived at the homestead, but what if someone else was there?

Both hands clenched the steering wheel as he pulled into MacDougall's drive. Lucille bucked like a rodeo bull in his rush to get over the ruts and humps of the lane. The lodge materialized as he lifted over the crest; a line of shadow from the forest dividing sunlight between the first and second stories. Malcolm wheeled in front of the porch and looked to the tree. The chain lay on the ground, lacking a dog.

"Perfect," he said tapping his fingers on the wheel. "I knew this was a bad idea." Malcolm gazed at the house. "Maybe he's inside." He opened the door and got out of the seat, walking on eggshells.

With each step onto the porch stairs he hesitated and listened. Reaching the top step, he approached the door, feeling as if he was penetrating an unseen perimeter. Why bother being quiet, if the purpose is to stir up the bastard inside? Malcolm thumped on the door with his fist. Only crickets and the evensong of birds responded. Sighing, he descended the steps and strolled onto the weedy lawn. Empty yard, empty porch, empty gloomy house.

Tall maples and oak on the western edge of the clearing obscured the setting sun. The grass swished under his tattered running shoes as he meandered into the back yard. Glancing at the small barn, the steers appeared uncaring of his presence bored in the pasture beyond, keeping company with a rusting hay rake. He went up to a back window, cupped his hands, and looked into a cluttered kitchen. A spent whisky bottle lay on its side on a dirty checkered table cloth.

"Damn it, this is a waste of time."

Malcolm stepped away from the window, resolved to either endure the mental anxiety of returning to this place again, or give up his mission entirely.

He cast another glimpse at the steers and saw an oblong stone in front of the pasture fence. Fighting his instinct to do otherwise, he drifted closer towards the marker and stumbled on a tablet set flat in the ground, 'Jebediah John MacDougall, 1886-1937.' He had infiltrated a ring of crumbling stones, graves hidden in the overgrown grass.

A chill nudged his neck, a reminder of the setting sun and the spreading shadow about him. The birds ceased singing their evensong. A lone vocalist creaked at the edge of the forest. "*Ka-ty-did, ka-ty-did,*" as a breeze murmured through the trees. Malcolm had overstayed his visit. He back-pedaled away from the gravesites, the light in the clearing weakening into a pallid aura. .

Pacing towards the rear of the lodge, a shape in an upstairs window caught his attention. He assured himself. Nothing more than light shifting off the panes. Keeping his eyes on the lodge, Malcolm quickened his gait. Lord, I'm farther away from the Scout than I thought. Other night screechers joined the soloist, chanting, "*get-out-now, get-out-now.*" With a sudden rush, crows cawed and dispersed from the edge of the clearing. Malcolm scrambled away from the stone relics into a dead run. Vague malevolence became physical reality as a guttural snarl, familiar to his ears, erupted from beyond the barn.

Malcolm peered over his shoulder, the black dog blasted from the shadowy woods, less than a hundred yards away. His pulse throbbed on either side of his neck as the trailing fury gained ground. Lungs heaving, he pumped faster, but the sound of the beast's panting increased with each stride. Malcolm resisted looking back, expecting to be grabbed at any moment. He veered towards a plastic lawn chair under a broad maple. With one last bound, he stretched his flying foot for the front edge of the chair.

Driving his leg off the chair, Malcolm maintained his upward momentum and grabbed a branch with both arms. The dog collided into the chair with unchecked speed, somersaulting in a duo of animal limbs and lawn furniture. Malcolm rolled his legs around the branch and hugged his body close to the rough bark. The beast recovered from its misguided inertia and sailed in the air underneath him with jaws that lashed a foot beneath Malcolm's hunched back. With a grunt, Malcolm inched one thigh over the branch, then pulled his torso on top. The dog lunged again, its front legs boomeranging off the tree trunk trying to extend the reach of its fangs. Malcolm wiggled his center of gravity above the tree limb and straddled the branch between his thighs.

Positioning the soles of his feet beneath him and extending trembling legs, he placed a hand on the trunk and stood up. His lungs dilated and shrank like a blacksmith's bellows with each gulp of air. Heart pounding, he closed his eyes to release his fear and catch his breath. He looked down upon the dog, which came to realize Malcolm was out of reach—for now. The creature settled down by the base of the tree. Yellow eyes regarded Malcolm, unmistakably as predator to prey.

Malcolm gazed at the Scout, sixty feet away, but might as well be six hundred. The katydids kept chatting in the dusk, scolding Malcolm for undertaking his fool's errand.

CHAPTER XX

John MacDougall dragged on his smoke. He casually looked to his left in time to see his pain-in-the-ass dutchman neighbor herding his sheep out of the pasture. To hell with the dutchman's sheep, nothing but a bunch of worthless animals and a waste of good grazing land. With one hand on the wheel, he rolled his truck off Shaneytown Road. A hundred yards ahead, Jack Calhoun was parked in the farm lane, sporting a shit-eating grin. MacDougall's eyes glowered at the occupant of the truck as he passed by. He wasn't intimidated by the Calhoun boys, no matter what they might think. His .38 revolver had a steady residence under his seat. He was fed up with their interference in his life.

Glancing in his mirror at the next crossroad, he lazily cut the wheel to the left. Didn't know what those Calhouns were up to, but he would avoid town and head for Cap's market instead. Cap stocked plenty of cigarettes and decent venison jerky. Besides, the shorter trip would get him back home in time to catch the start of the ball game on TV.

*****

That MacDougall would take Malcolm's presence in the maple tree unkindly was an understatement. A simple arrest for trespassing would be a blessing. The sun was his hourglass of reckoning, slipping towards the horizon and marooning Malcolm in the gathering twilight. The overbearing shadows that crept out of the woods gnawed at his nerves. A breeze hissed from the direction of the lodge and the leaves in the tree trembled about him.

A premonition crossed his mind to avoid looking at the dwelling.

I should have stayed with the plan and not wander away from Lucille. His folly became fully apparent. While on his way to Dale's farm lane, Malcolm hadn't passed MacDougall's oncoming truck heading for town. Dale and Jack may not have been able to tail the white truck after all. In desperation, Malcolm weighed the only other option, to drop down on the canine thug below and take his chances.

Crickets and katydids metered the time that remained before Malcolm's judgment. In a spruce tree on the far side of the lodge, a mourning dove lamented his plight. Less than ten minutes had passed since he scrambled up the tree, it felt like hours. An engine groaned in the distance on the road. Malcolm listened, not knowing if he wished the sound to continue past, or turn into the lane to cease the terrible dread. Even from nearly a hundred yards away, the change in engine pitch signaled the vehicle was slowing down. The black dog stood up, but remained by the tree. Malcolm inhaled, but failed to release the tension around his girth.

The whine of a truck's transmission sounded on the lane. Although unseen over the rise, the glow of the headlights grew stronger. Malcolm thought, damn it, here we go. At least he got his wish to evaluate the behavior of the dog, no doubt unchanged. Jonas P. wouldn't have to worry about rabies after all. The glare of the beams broke over the rise and the pickup came into view.

As it swayed into the clearing, the truck veered past the old coupe away from the lodge and towards the tree. The dog barked violently and leapt at the passenger window of the truck. A voice yelled, "Git out a here, you furry son of a bitch!" Nearly clipped by the front bumper, the dog recoiled out of the way of a *blue* truck.

The pickup halted underneath the tree. Malcolm squinted against the glare of a flashlight on his face. "Saints be praised, Doc, you got yourself into a bit of a jam." Malcolm squatted onto the limb, trying to discern the voice stemming from within the brilliant daze. "Well, if you've had enough of playing the monkey, jump on down into the bed and Jimmie here'll chauffeur you over to your

Scout." Malcolm didn't need to be asked twice. He hung from the limb and let go.

Malcolm's shoulder banged into a wheel well as Jimmie spun in a tight circle. He gained his knees and peered over the side of the bed. The truck sped up to the driver's side of the Scout, followed by a sharp crack as the side mirror snapped off. Shannon's head poked out from the passenger window of the truck. "Doc, climb in through your window!"

Malcolm crawled over the wall of the truck bed and burrowed head first into Lucille. Twisting and turning his legs in the tight space, he got behind the wheel and started the engine. Jimmie sped towards the lane, the black dog at their side, relentlessly goaded by Shannon. A shiny glint caught Malcolm's eye. On the ground lay what remained of the mirror. Damn it. To leave it there would offer MacDougall a calling card of the ill-fated visit. Jimmie's truck vanished over the rise, the dog still in pursuit.

Malcolm popped out of Lucille to fetch the mirror. He collected the bracket off the ground and fingered about the soil to find the shards of the mirror. He froze at the sound of a voice chattering behind him. Whirling about, he was stunned to see a middle-aged woman, dressed in a Victorian-styled nightshirt swaying in a rocking chair on the porch of the lodge. Her barren eyes stared into the woods as she hummed a tune. He leaned back slowly, his feet glued to the ground, unwilling to create any movement that would attract the singer's attention.

A belligerent growl from the direction of the lane shook his focus and for a second time he found himself in a foot race against his nemesis. He plunged into the seat and yanked the handle, two seconds before the momentum of the brute slammed against the door and secured the latch. He threw what pieces of the mirror he had onto the floor and Lucille lurched into gear. The beast sprung at the open window. Malcolm flinched and almost grazed a tree. He swerved back on course while his pursuer assaulted his left front tire.

The Scout ripped out of the lane and cut a sharp left onto the road but the beast refused to relinquish the chase for at least another fifty yards. Careening along without a clear view of the pavement, Malcolm turned on his headlights with an unsteady hand. He sped by Jonas P. Yoder's place and almost reached the "Y" when an oncoming white pickup passed him.

Taking the turn back towards town, a flashlight waved him down. Dale stuck his head into his window. "Doc, you better come down to my place and steady yourself a little while."

Dale slid into the passenger seat and Malcolm rolled down the lane. As Malcolm stepped into the barnyard, Shannon materialized with a quart mason jar that smelled of paint thinner. "Here Doc, I think you need this." Malcolm sniffed the elixir and took a sip. Fire spread down his throat and into his stomach. He added another, longer sip.

Malcolm passed the jar to Dale. "That was pretty foolish."

Shannon asked, "Doc, how *did* you get yourself treed by that son of a bitch?"

Tautness melted into numbness throughout Malcolm's body. "I didn't see him when I pulled in, thought to check the house, wandered about in back, and it burst of the woods like a demon from hell. I was nearly caught."

Taking the jar from Dale, Jack stated, "Someday that evil bastard will be evil no more."

Malcolm thought out loud. "Well, at least I know he's had no behavioral change since getting after Jonas P.'s sheep." Malcolm collected himself for another minute while the jar was passed around. "I owe you one Shannon, and you too Jimmie. How in the world did you know about my predicament?"

Shannon lit a cigarette, paused to enjoy the delicious intake of the nicotine buzz, and blew smoke up into the night air. He waved the cigarette between fore and middle finger towards Malcolm. "Well Doc, we were the backup you see. Dale and Jack, they were to keep an eye on ol' MacDougall. We were just up the road from here. It turns out that MacDougall didn't go into town, might have gone over to Cap's EZ Mart. Dale and Jack couldn't find him, so they radioed us on the shortwave. We figured we better make sure you got the hell out of there." Shannon let go a raspy cackle. "We sure didn't expect to see you up in no tree." Laughter rippled around the circle.

Shannon collected himself. "You know Doc, you're one smart fella, all that book learning. But you need a little teaching about local issues if you know what I mean."

Malcolm said, "I'm glad you did." He took a close look into the jar in his hand. "What is this turpentine we're drinking anyway?"

"It's a little Amish firewater, Amos M. makes it."

Malcolm furrowed his brow. "Amos M. Yoder?"

"Jonas P.'s brother." Shannon rested his free hand on Jimmie's shoulder. "He won't see this on no farm lane sign along with the veggies, will he Jimmie?" Jimmie shook his head, grinning. Shannon went on, "Yeah, a few of the white toppers make a little white lightning, for medicinal purposes only mind you." Dale slapped Malcolm between his shoulder blades. "And no Sunday sales."

CHAPTER XXI

Malcolm sagged into the cavernous sofa, resonating a midrange D with each flick of his fingernail on the green glass of the beer bottle. His host returned with two cold bottles, one of which ended up in Malcolm's hands as a match for the empty one. Bobby dropped into an arm chair on the opposite side of the room. "Still fretting?"

Malcolm shrugged. "Maybe I'm not the socially engaging type."

"It's time to move on, man."

"To where?"

Bobby swigged his beer. "I've been through the divorce dance, too, you know."

"And?"

Bobby held up two fingers. "Two simple rules, seek women who are drivers and don't go where you aren't wanted."

"That's profound."

"Think of it this way, some folks are afraid to drive down roads they don't know. Others will travel to new places, but only as passengers." Bobby leaned forward and rested his forearms on his lap. "But then there are the drivers, who aren't afraid to hit the accelerator." Bobby took another slug from his bottle. "Of course, you gotta be careful with someone who *never* uses the brakes."

Malcolm rubbed a temple with his fingertips. "And this barnyard philosophy applies to me how?"

"Your ex didn't want to travel away from what she knew and sure as hell didn't want to be your passenger."

Malcolm sighed, "Not for a ride in Galloway County, anyway…and the 'going where wanted' part?"

Bobby leaned back. "That's all you. Forget about forcing your way back into a relationship where you're not welcome. You should be looking for someone more inviting."

"Like Katy the barmaid?"

With a push of the door, a small cyclone burst into the scene, greeted Malcolm, hugged Bobby, and blew into the kitchen to find a beer. She called out, "Have a good week, Malcolm?"

"It was okay."

"Ready for a good time tonight?"

There were times when Malcolm thought Marcy Snedeker was a bit too enthusiastic about life. The present situation was one of them.

Bobby said, "If we have the same band as last year, the music will be good."

Marcy plopped herself on Bobby's lap. "The caller gets everyone involved and picks up the pace as the night goes on."

Malcolm asked, "Is there a novice section?"

"We'll have to find you a reliable partner."

"She needs to be quick to avoid my stepping on her feet."

"You couldn't do any worse than Bobby last year."

"That's comforting." Malcolm's mind drifted to his last dancing performance...his wedding night, an orchestrated soirée of tuxedos, gowns, and protocol, so much protocol. Carrie had been in her element as she gracefully mingled and entertained.

"Malcolm, are you with us?" Marcy and Bobby were laughing.

Yanked from the past, he lowered his eyes. "Sorry, just thinking of previous events."

Marcy's said pedantically. "It's high time you let go a little."

Malcolm nodded. "I'm ready."

"Alright then, let's go."

Bobby lingered in his seat, pointed to Marcy's back as she was halfway out the door and whispered, "Driver."

Malcolm followed Bobby's truck on a road that bisected fields of alfalfa and emerald corn. The trip came to an end in a clipped hayfield serving as a parking lot. Laughter and conversation resonated through the warm air, seasoned with the smoky drift of barbeque. Marcy ricocheted her way through several groups of people, eventually finding one of the hosts. "Dan Gilbert, I want you to meet Malcolm, our friend we told you about."

"How's it going?" beamed Dan.

"Thanks for the hospitality." Malcolm shook his hand, "I hope you don't mind if I brought along a little single malt for the occasion."

Dan appraised the rectangular box. "Brother Dave will be happy to see that. Make yourself at home, the pig is ready for pickin', we've got roasted corn and potatoes and all sorts of stuff that people keep bringing. Keg is in the far side of the barn. The band is setting up as we speak."

Marcy said, "You know us. We'll find what we need." She raised herself onto her toes and scanned the throng. "Bobby, will you find a beer for me? I'll be right back."

As she streamed through the crowded barn, Bobby said, "I guess we'll be forced to get one for ourselves as well."

Malcolm agreed. "I think we can handle that." They stopped by the barbeque and subsequently befriended others who had also discovered the location of the keg. Marcy correctly determined the most likely spot to find them. She wasn't alone. Marcy asked, "Malcolm, I believe you may know Jolene?"

Malcolm gulped his mouthful of makeshift pork sandwich. "Yes, I've had the pleasure previously."

Jolene replied, "Good evening doctor, I hear you're quite the barn dancer."

"That's a bit of an exaggeration." Malcolm circled his hand in front of the group. "You all know one another?"

Marcy laughed, "Jolene and I are best friends."

Malcolm flashed a sideways peek at Bobby. "Well, isn't that a coincidence." Bobby handed his beer to Marcy and took to filling up two plastic cups at the keg with conspicuous concentration.

Malcolm said, "I take it you've been to this event before, Jolene?"

"Oh, a time or two."

Bobby raised his eyes from the tap long enough to wink at Malcolm and silently whisper, "Driver."

Marcy added, "Jolene doesn't get out much. You know these farm girls. All those chores."

"As I recall, a very capable milker," Malcolm said.

"I fill in once in a while. I think my brother was taking a week off during your last sightseeing tour of the farm."

Malcolm thought that Marcy giggled a little too easily at this comment. "And how about you Marcy, ever help Jolene milk? It's a good life skill if you ever get that backyard goat."

"I go there to keep Jolene company. I like feeding the calves."

Jolene said, "But we do have to watch for tourists who stumble over things that get in the way." Marcy put her hand over mouth and chortled. Bobby returned with two cups spilling foam over the brim.

"I *knew* you were good for something," Marcy exclaimed.

Malcolm inquired, "How about you Bobby, many gawking tourists come to visit your farm?"

"Gawking tourists? No. But future graduates from the county charm school, yes."

A guitar on the other side of the barn plucked a rhythm. A voice descended from a speaker tied to the ceiling beams, inviting all to gather onto the worn plank floor. Marcy grabbed Bobby's arm. "Let the good times roll!" Malcolm offered a hand to Jolene. She measured the length of his arm from hand to shoulder with her eyes, then snaked her arm about his. A current flowed in the core of his body. A little later than sooner, the caller prevailed in gaining some semblance of order among the jostling assembly and a fiddle sang out an easy tempered reel.

Armed with a quick wit, and despite the subdued lighting, the caller targeted dancers who strayed like flotsam away from the main flow. Try as he might to follow Jolene's lead, Malcolm was the center of attention with regularity, transcended from reality with each clasp of her hand and touch of her body. The band stepped up the rhythm and the vortex spun faster.

The first set ended and the foursome enjoyed sudsy beers outside a barn door. Tentative stars mustered into an irregular array as day softened to twilight. Malcolm struggled to focus on the conversation as Jolene's hand reached about his waist. Time stood still, until the music summoned everyone back into the heat of the barn.

Marcy and Bobby attached themselves to the swarm. Jolene stalled. "Want to take a walk?"

They threaded their way against the current, towards the cool anonymity of a small pasture, refraining from speaking as they strolled into the edge of darkness. The pungent odor of pot wafted in the air. Memories of college sifted through Malcolm's thoughts as he noted a cluster of tokers.

Coming across a hay wagon, Jolene hoisted herself onto the worn oak boards. She rested her back against the bales and dangled her legs over the edge. Malcolm sprang onto the wagon and settled next to her. The clamor from the barn dissolved within the cloister of fresh cut alfalfa.

She mused, "Must have been a culture shock to come to this county."

"I have to say it's a rather unique place."

"Whatever possessed you to come here?"

"Hmmm, I'm not sure." His mind reset to age sixteen and his brand new driver's license. His car privileges were often limited to waiting for his dad behind the wheel of a Pontiac station wagon, under the musty glow of mercury vapor lamps in the parking lot of the train station. Green and yellow double-decker trains would roll up and spew scores of suit-wearing penguins out the doors, sprinting to faithful spouses in strategically parked cars. A scrum from the parking lot back to the highway would ensue; office moxie continued in order to grasp an advantageous spot in the traffic line. From there, the drive home for dinner and a night of mind-sapping TV. Comic relief to release the traumas of overbearing bosses, neighbors who had a better car, and above all, the angst of being branded a nonconformist.

His eyes set upon Venus, a silver diadem in the space recently relinquished by the sun. "I have no explanation for my being here, given where I was raised, but I like working with cows and I like working with farmers. They can be a bit gruff, but you know where they stand."

"Knowing where my dad stands on things isn't much of a problem."

"That's for sure, especially all of those home remedies."

"You disapprove?"

He tried to read her face in the deepening shadow. "If he wants to believe in them, good for him. I worry he's having his money taken from him, that's all." Malcolm revisited his scan of Venus, provocatively concealing herself in a veil of amethyst on the horizon. "What's it like to be an insider from Galloway County?"

"People here are happy with the status quo. Boys want to grow up to work in the mill and like their dads; and the girls get pregnant and beat up by their drunk husbands." She paused. "I needed to get away for a bit."

"But you came back."

"Dad hurt his back falling from a silo a year ago. I figured to return and help out a little. My brother Marty will probably take over the farm, but they could still use a little help. I was able to get a job editing for the *News-Gazette*, at least put my college writing to use."

"So you have an altruistic streak."

"Or martyr."

She leaned towards him, reached around his neck and pressed her lips to his. His hands clasped the small of her back and the physical world about them faded into starlit solitude. Lowering a hand across his shoulder, her fingers stroked his chest and softly pushed away. "I've got to go. I promised to help with the calves tomorrow morning. Mom and Dad want to get an early start to go visit her sister."

Malcolm exhaled. "Alright, I should probably wander back,

too." She embraced him as they hopped off the wagon. His forearms rested on her bare shoulders.

"Why don't you call me so we can continue our conversation? You can get my number from Marcy or Bobby."

"I'll take you up on that offer."

She touched his lips with her index finger and they returned to the world of brazen light and sound. Arriving at the barn door, she caressed his cheek, spun through the portal and blended into the pack with the grace of a deer in a forest.

Dan Gilbert shouted out, "There you are! Bobby says it's time for you, me, and Dave to enter into a philosophic discourse."

"Well, I was thinking of slipping on home..."

"Can't 'slip away', until you say you're going to slip away at least three times." Malcolm didn't offer much resistance. He was more at ease than he had been for a long time. Divine lightning had struck, leaving remnants of charged particles coursing through his veins and empowering his fingers to ignite objects by touch.

A waning crescent moon was rising as Lucille found the way home. Anxious barking from within the house greeted Malcolm as he lumbered out from behind the wheel. "Okay, okay, I'm coming," he called out. As he tugged the front screen door open, a small cardboard box rolled onto his feet. "What the hell?" There were no markings or wrapping. He picked up the package and pushed inside. Precious ran in a circle by the back door.

"Sorry, I stayed later than I planned." Malcolm turned the knob; Precious popped past the screen door and was gone. Malcolm sauntered into the kitchen, tapped the light switch and popped opened the box flaps. His newly acquired life force sank through his feet and onto the floor. Small fragments reflected light from within the box. He picked up a scrap of paper that accompanied the glass collection. "Next time, I'll make goddam sure you won't get away from the dog."

FALL

CHAPTER XXII

Squeezing pieces of side mirror in his hand, vintage 1976, his ears resonated from a diatribe about veterinarians, particularly young ones, who drive as if possessed by demons—or trying to elude them, Malcolm thought—jam their vehicles into preposterous places—where they didn't belong—and have so little respect for a critical part of their livelihood.

"Jeez-oh-Pete Doc, we *just* found that damn mirror but a week or two ago, and here it is, knocked right off. A drunken chimpanzee could steer your car better than you."

Compliments of ragweed pollen, Malcolm rolled a finger across an itchy eye, unwilling to remind Floyd that the previous mirror replacement was more than two months ago.

A grimy hand thrust out at him. "Alright, give it to me then."

Malcolm sheepishly surrendered the fragments. Disgusted with the exhibit of careless auto ownership, Floyd smoothed his thinning grey hair. "Jeez-oh-Pete."

A carbon copy of Floyd's head, plus three additional years of hair loss, protruded past the doorframe. "Tougher to find parts for those IH Scouts every day."

"Yupper, Lloyd, gets tougher every day."

"Don't make them anymore."

"No sir, they don't."

"Should be more careful when driving."

"Yes indeed, Doc should."

"Take a while to find that part again."

"At least a week."

"Cost some money."

"Sweet Jesus yes, more than last time."

Malcolm twitched. Tweedledee and Tweedledum in stereo.

Floyd removed his glasses to peer closely at the debris. "I think it's fixable, but the screws got all bent out of shape. You're going to have to go to the glass shop in Gillstown to get a new mirror cut. What in God's name did you run into this time?"

Malcolm sniffed his runny nose and considered his reply. "A pick-up grazed me in a rather tight spot."

"You know Doc, I've seen demolition derby wrecks cared for better than your Scout." Floyd cleared the clutter from a few square inches of his desk to rest the mirror. "Criminently!"

"While you're at it, any chance you can try to fix that ignition again?"

"If I was your Scout, I wouldn't want to start, just sit in the garage."

Lloyd nodded. "A regular ol' Richard Petty."

"When was the last time you checked the water level in your battery, Doc? You know you have to stick your head under the hood once in a while, especially a guy like you who lives in his vehicle, which by the way could stand to be cleaned a little better."

Malcolm slowly shifted out the door, muttering silently, "Yeah, yeah, and eat all your spinach, and mow the lawn, et cetera, et cetera." Still recovering from the harangue at the gas station, Malcolm checked in with his other surrogate parent on the radio.

Lois queried, "Is it true Malcolm, did you lose your mirror again?"

Judas Priest, even she knows? "Yes, Lois."

Another pause. "How's Jolene Brubaker doing?"

Malcolm sneezed. "I'm on my"—sneezed again—"way to Mahlon R. Zook."

"Let me know how things go." He clipped the microphone back onto the dash. Probably half the county knew about his barn dance liaison and the other half knew about the missing mirror.

The morning fog had finally melted into a mellow, hazy light.

Monarchs flitted about milkweed and yellow spikes of goldenrod. Lingering dew garnished sagging spider webs with iridescent beads. The valley was consumed in the all-out annual assault to ensile the corn. Chopper blades slashed ranks of maize conscripts down at their knees, their shredded yellow-green remains hauled away in wagons for mass burial in deep silos.

This was the season when Amish kids sprouted up along the roadsides, clustered together as they walked to white schoolhouses, landscaped with two outhouses, a buggy parked in the yard, and a horse tied under a tree—the teacher's commute home.

At the end of the day, Malcolm pondered his garden over grilled chicken. Sadly, the sweet corn was past season, but the last of the fresh tomatoes were still available. He paced on the patio, the smoke of the grill roiling about him; it was time to summon his nerve to call her.

*****

Jolene hung her legs over the armrest of the glider. She liked the vibe of this small house and the rent was cheap. Perched on a hill, her front porch overlooked the Gillstown town square, the metal roofs gleaming with an apricot glaze as the sun lowered in the west. The windows of many of the stores and businesses she remembered from growing up were boarded, devoured by the cavernous box stores on the edge of town. But a great diner for breakfast still survived and the *News-Gazette* was but a short walk away. Her journalism career revolved around local school and township board meetings, kudos for the dairy farmer with the highest producing cow in the county each month, elections for volunteer fire company officers, and high school sports scores. She could have written the same stories ten years ago, the names and events hadn't changed much; such was life in Galloway County.

Jolene stretched out and clasped her hands behind the back of her head and contemplated Malcolm. Marcy and Bobby said he was in need of someone to help him navigate local culture. A 4x4 pickup, jacked up on a high suspension and in need of a less obtrusive exhaust system, roared by on the street. She grimaced as her eye caught the rebel flag on the back window. A reminder of how easily the local culture could vex her at times.

After their encounters at Delmar's farm and the barn dance, she believed he was genuine and had a sense of humor, if somewhat sardonic. Her dad thought he had the makings to be a good vet. No doubt he was smart. Why was he here, in Galloway County? She perceived an untapped energy within him. She would have to pry on Marcy to find out more details about his divorce.

She sat upright, a premonition twitching the back of her neck. Moments later the phone rang. Sliding through her screen door, she picked up the receiver and asked confidently, "Finally got the number?"

"How did you know it was me?"

"The timing was right."

"I've been busy saving lives and stamping out plagues."

"Really?"

Malcolm paused. "Your week going well?"

"Yeah, another dose of the edgy daily life of metropolitan Gillstown."

"How's Delmar?"

She pictured him winding the phone cord in his free hand. "Still Delmar."

"Um, I have this weekend off. Would you like to go out to dinner?"

She sensed the cord fidgeting was becoming more animated. He doesn't like talking on phones. "That sounds good, have you ever been to Taggart's?

"Taggart's?"

"County landmark, it'll round out your experience in the valley."

"Valley, as in my side and not the Gillstown side of the mountain?"

She thought. He is so out of the loop about so many things

around here. "Yeah, we'll have to wait in line maybe. A bring your own beverage place, I'll get a bottle of wine."

"What do they serve?"

"Fried chicken, waffles, syrup, green beans, and coleslaw."

"What else?"

"Fried chicken, waffles, syrup, green beans, and coleslaw."

"Must be done right to be so popular."

"You can be the judge. I'll pick you up at your place on Friday at about 7:00. No sense you driving here and then back into the valley."

"Need directions?"

"I know where you are."

"How?"

"Everyone around here knows that."

"Since when did I become such a celebrity?"

"It's more notoriety rather than celebrity. You're not from around here, are you?"

"Very funny. I'll see you on Friday, then. I'm looking forward to a gourmet experience in the valley."

"Good night, Malcolm."

"Good night, Jolene." He hung up. She nearly laughed out loud. How long would it be before he finally caught on about how easily news spread around here, even when it was little more than idle gossip?

CHAPTER XXIII

Jonas P. Yoder strolled through the long shadows of early evening, trailing a vanguard of swirling fleece. At his side, Kip and Fritz divided their attention between the flock and their master's body language. A young ram stamped his feet and refused to enter through the gate. With a whistle, Jonas P. directed Kip to rally the ram to join the common cause. A blur of black and white fur flashed in front of the ram's nose. The startled sheep retreated to the flock, only to be pursued and pulled onto the ground by the back leg. "Kip! KIP! *Was machst du?* I don't need that kind of help!" Eyes smoldering under the brim of his hat, Jonas P. pointed to the barn. "*Gehst du!*"

Kip crept under the fence, tail low and ears down. Jonas P. got the last of the sheep behind the gate and slid the latch.

Standing by the barn, Rachel clapped her hands and softly consoled the outcast. "Kip! Kip! *Komm hier!*" Relieved to find a friendly voice, the dog trotted over to the girl. "Good dog. A little too much, *ja?*" She crouched to touch her nose to his. "Good dog." Kip responded by licking her face.

*****

Gluttony was a sin and Malcolm was in need of absolution. The chicken and fritter feast kept appearing at Taggart's family style dining until all the diners at the shared table could no longer stand the sight of another drumstick or wing. He leaned back in the seat of Jolene's truck, wishing he could recline a bit farther.

She stole a glance from the corner of her eye. "You can really put the food down.”

“Fried chicken brings out my compulsive behavior.”

“We like to encourage newcomers to frequently eat at Taggart’s.” She poked his distended abdomen. “Fatten them up to get them suitable for roasting during our pagan harvest rituals.”

Pondering his misguided rationale for eating a second piece of apple pie, Malcolm turned his head towards her. “How many victims are usually required for the sacrifice?”

A smile lit her face. “Usually one, if of adequate size.”

“Does that event get covered by the *News-Gazette*?”

“In the community events calendar, as part of the church news section.”

“So where did you get your training to be a journalist?”

“I always liked writing in school, attended Slippery Rock State.”

“Finish your degree?”

“No, got a little restless at the end of my third year, needed to leave and see another part of the world.”

“Classroom boredom didn’t appeal to you?”

Her hands tightened on the wheel. “Let’s just say I had enough of school at that point in time.”

Not wanting to blunder into a minefield, Malcolm feigned interest in the amorphous shapes and shadows flashing by the passenger window.

Jolene rolled her head to release tension in her neck. Strands of hair tumbled over her shoulders; gravitating his energy towards her with no more effort than the sun harnessing a comet into orbit. “Where did you get that antique Scout?”

“Give her some respect, she’s only eight.”

“She?”

“Her name is Lucille.”

“Where did you come up with that?”

“Shouldn’t give up all my secrets on the first date.”

“So you’ll probably hang on until the wheels fall off?”

His teeth flashed in the dim light of the cab. “I’ll bet on the transmission first, I can take you on a spin if you’re that interested.”

"No thanks, I'll stay with my truck. And why the black paint spots?"

"The original color was white, got into a couple of fender benders. I sprayed on primer thinking I would eventually refinish painting the entire body. Never did, decided I liked black spots rather than gray."

"Professionally done."

"Consider it a moving piece of art."

"This is not the kind of place where art is much appreciated." She reached out her hand and touched his cheek. "I have an idea for next weekend, how about a hike in the ridges? The leaves should be turning color."

"Can I take Precious? She hasn't been out for a good long walk in a while."

"Your dog?"

"She's a scrappy little beggar."

She pulled the truck into his driveway. The reflection of the headlights against the garage door revealed their faces. She leaned over and kissed him then rested her fingers on his chin. "She can be our chaperone."

"Implying we'll need one?"

"Saturday morning then." She paused. "You know what? Let's make the occasion a double date, I'll take along Pierre."

"Pierre, the mad English Sheepdog?"

"I'm sure he'll get along with Precious, he's harmless."

"If not a little over-amorous."

"He's neutered."

Malcolm refrained from noting the benefits of limiting genetic dispersion by certain individuals.

CHAPTER XXIV

Small stands of withered corn, the rear guard of the growing season, shuddered in their khaki uniforms against the assault of wind and rain plunging over the ridges. Chased by the wind, leaves scattered past Lucille's windshield like a frantic flock of red-yellow birds. Malcolm came upon a file of yellow hazard lights blinking in offbeat rhythm in the opposite lane. Unable to discern the occupants through the rain-beaded glass, he swept by the lead buggy and raised his hand in salute. The wedding season had arrived in the valley—Tuesdays and Thursdays for an entire month, a fall festival of family, fellowship, and fecundity. Malcolm couldn't fathom the religious or superstitious significance of those particular days of the week. Part of the nuptial experience included the newlyweds briefly lodging with relatives in an elaborate sequence of family hierarchy. First the respective in-laws, followed by aunts and uncles, cousins and perhaps married siblings. Malcolm envisioned honeymooning with *his* in-laws. It was a disturbing thought.

Lucille splashed into the muddy wash of Nahum Y. Peachey's barn lot. A pale yellow light glimmered through the windows of the barn. A figure appeared with a kerosene lamp in the open stable door. Malcolm rolled down his window as Nahum Y. peered into the Scout, his face cast in a ghoulish mask by the lantern. "How are you, veterinary? A stormy evening, *ja?*"

Malcolm's eyes heeded the weaving branches of the sturdy oak by the milk house. "Indeed. Where is the cow?"

"In the pasture, *komm.*" Nahum Y. sloshed across the barn lot, rain pelting off his black denim coat. Malcolm muscled Lucille's door open against the wind and locked the front wheel hubs into four-wheel drive. Nahum Y. endured the torrent to keep the gate open as Malcolm rolled into the pasture. Trailing the farmer down a gentle slope to the left, the field leveled off by a small stream. The rain dissipated to a fine drizzle that scintillated in the beam of the headlights. Malcolm spied an ill-defined mass that he took to be his patient. Her back end was in the stream. "Location, location, location," he murmured while he gathered his supplies, not bothering with the bucket.

The rain ceased amid a damp wind that fragmented the clouds and buffeted Malcolm's back.

"When was she fresh, Nahum?"

"Yesterday morning."

Malcolm gave the cow a quick physical and pulled a bottle of calcium from a pocket. "Did she clean?"

"*Ja.*"

Finishing the infusion, Malcolm pulled the needle and urged the cow to get to her feet. She remained uninspired by his efforts. Malcolm repeated the prodding, with no change in her effort. Nahum Y. yanked on the halter while Malcolm tapped her back with a needle. This only incited the cow to moan and shift her weight from one side to the other. Bovine silhouettes approached the creek from the slope of the pasture. The down cow's ears perked up and she called to her herd mates in a low hum. Malcolm continued to whistle, yell and futilely push against the fifteen hundred pound mass for several minutes. He sighed, "Nahum, you might need the horse team to pull her out of the stream. I'm afraid she may lose her body heat lying here all night." No sooner had he finished speaking when the rain returned and lashed his face with a vengeance. At times like this he almost thought he would have been better off to return to Illinois. Almost, but not quite.

Foiled by the cow and the weather, Malcolm trudged out of the bone-chilling creek towards Lucille. The herd drifted up the rise to keep their distance from the unknown human. Ardent grunts and splashes followed Malcolm. He turned to see his patient gain her feet and stagger out of the water. Malcolm and Nahum Y. looked at one another. "I guess maybe she wanted to be with her friends, so she did," Nahum Y. said.

Malcolm weighed in on the mystical process of the bovine mind. "I suppose so." The cow waddled into the evening, her inertia overcome by the power of her herding instinct.

"I'll meet you at the milk house, after I close the gate," Nahum Y. said, leaning against the gusts of wind.

Nahum Y.'s lantern cast a primal comfort in the milk house as the pelting rain battered the metal roof. Malcolm queried, "Who had the wedding today?"

"Kore J. Byler of Jericho Road, his son, Mahlon, and John R. Peachey's daughter, Sarah."

"Too bad the weather was so miserable."

Nahum Y. shrugged. "The rain comes when it comes. That's what barns are for."

"Is Mahlon going to farm?"

"I think he'll work at his daddy-in-law's farm for now."

Malcolm pounded his boots with a soapy brush. "Farmers have all these weddings as an excuse to fatten up for next year's field work."

Nahum Y. beamed. "We had turkey, ham, turnips, squash, potatoes, beans, chow-chow, applesauce, *und Apfelkuchen, mit Vanille Eis.*"

"Apple pie with ice cream? That's a lot better than I ate today." Malcolm dumped the water from his bucket into the floor drain. "Call me tomorrow morning if the cow has any problems, Nahum. Have a good evening."

"*Guten Abend, Tierarzt.*"

Thunder echoed off the walls of the milk house. It wasn't from the clouds above. Malcolm cocked his head. "Your bull?"

Nahum Y. furrowed his brow. A strong gust of wind flung the door open and the gaunt form of Zephaniah Y. Peachey materialized in the opening. His son lowered his head.

"Good evening, Mr. Peacher," Malcolm said. "Pretty wild night with the wind and rain."

The elder man anchored a one-eyed jab on the young veterinarian. The bellowing beast in the stable resounded with such intensity that it seemed as if he was right outside of the milk house door.

Dialogue on the part of the elder was not forthcoming, which relieved and inspired Malcolm to collect his gear. "I guess I'll be going now. *Guten Abend.*"

Zephaniah Y. shifted his weight onto his cane to allow Malcolm passage. Malcolm noticed Nahum Y. kept his eyes downcast. What a whack job he's got as a father. Bracing himself against the wind, Malcolm headed for the sanctuary of Lucille and home. His headlights glanced off the billboard at the end of the lane. **Therefore let us not sleep, as do others; but let us watch and be sober. For they that sleep, sleep in the night; and they that be drunken are drunken in the night. Thes 5:4-7.**

*****

Precious snored by the side of the bed, keeping time with the patter of raindrops on the windows. Malcolm dozed off, submerging into a subconscious journey across a stark landscape with no trees or grass. As he worked his way around jagged boulders, a flash of lightning sliced the air, shattering the rock in front of him. He fled, but the roar of thunder confronted him from every direction, as boulder after boulder disintegrated. He awakened short of breath. Precious was running about the house, overcome with rage.

Collecting his wits, he disentangled himself from his sheets and stumbled onto his feet. Precious was in one of the front rooms, that he used as his makeshift office. Malcolm ignored the light switch in the hall, groping his hands along the walls to steady his balance. He had about reached the doorway into the first room on the right when a sharp crack erupted from outside the house, followed by the sound of breaking glass. "Holy shit! That's a gun!"

He crouched down as his head inched past the door frame.

139

Precious was beside herself, howling and springing onto the window ledge with her front paws. Another shot rang out, this time cracking the window of the spare bedroom next door. Malcolm screamed, "Son of a bitch!" and lunged across the office. His thigh glanced off a corner of the desk as he sprawled over Precious and tackled her onto the floor. She squirmed in his grasp like a fish caught in a net, forcing him to tighten his clamp. "You stay down!"

An engine droned up the road and away from town. Precious continued to yip in an excited staccato at the window. Willfully, Malcolm slowed his respiration, refusing to move for several breaths. He released Precious, needing to free himself from her panting and body heat.

Irreparably awake, Malcolm collected his robe from the bedroom and poured a couple fingers of single malt in the kitchen, without the benefit of light. He collapsed in an easy chair by the picture window overlooking the garden. The rain had receded, leaving moonlight and a hint of a frost behind. Precious relaxed by his feet, lying on her side, legs stretched out. His preternatural awareness faded away, a welcome effect of twelve-year old scotch.

As Malcolm slid between the sheets for the second time that night, Precious rested her head on his bed covers. "You still seem a little wired." He patted his leg. "Okay, come on up." Before he could turn off the light, she was curled up on the foot of the bed. Judging by her lack of snoring, Precious, like Malcolm, slept lightly.

CHAPTER XXV

Brisk morning air and an orange tabby entered the back door, in exchange for Precious bolting out.

"For what do I owe this privilege?" The cat took little notice of Malcolm and stalked to a place of his liking on the couch. Malcolm started a pot of coffee and dropped bread into the toaster. Precious whined outside on the patio. He let her in, and she made her way under the kitchen table. "The toast will have to wait until I take a shower."

He poured coffee into a mug and padded down the hall in his bare feet. Coral light seared the windows of his makeshift office. Squinting into the rising sun, a glint caught his eye. Matching prisms radiated from the sashes of both windows. Malcolm edged across the room for a closer look—two holes, cratered on the inner surface of the glass. The hair on his arms lifted. "Judas Priest. Two shots here…not one."

Malcolm found identical fissures in both windows of the spare bedroom. Precious had joined him, raising her nose to gain a scent. He reached down and stroked her back. "Got ourselves into a bit of a mess, didn't we?"

His mind seething with possible motives and perpetrators, Malcolm finished his shower and tracked down his neglected coffee. Pacing between the kitchen walls, he cogitated on Jolene's arrival in the afternoon; maybe she could provide local context to the target practice. He buttered his toast and munched on a piece as he returned to his bedroom. Time to get dressed and sort things out from an outside perspective.

Under brilliant sunshine, Malcolm touched the edges of a hole in one of the windows with his index finger. A car climbed up the hill from town but didn't accelerate at the summit nor continue to the highway. Malcolm about-faced to watch a black and white Crown Victoria roll into the driveway and park behind his Scout. Cal Geisfelt exited the car and pushed his door shut.

"Morning, Doc", Cal amicably greeted, his eyes masked behind reflective aviator sunglasses. "Getting a little housework done?"

"Good morning, Chief, what brings you here?"

"Oh, it's a beautiful day for a drive in the neighborhood, I guess." Malcolm detected a slight limp in Cal's right leg as he came around the side of Lucille. Cal looked past Malcolm, took off his shades, whistled a few chords, stepped back and scanned the entire front of the house. "Doc, if I didn't know any better, I'd say that somebody put some .22 rounds through those windows." Prompted by Malcolm's silence, Cal suggested, "One might typically call the police on such occasions of reckless endangerment."

Malcolm folded his arm. "I guess I was trying to collect my thoughts. I would have called you after a bit."

Cal's flint colored eyes narrowed to slits. "Happen to have any idea who might have done this, or see anything?"

Cal's tone tickled a little part of Malcolm's brain to respond carefully. "I was home, Chief, had fallen asleep. Maybe 11:30 to twelve. I awoke to find the dog barking. I heard a vehicle speeding away after the shots were over."

"But you didn't see the vehicle?"

Malcolm pursed his lips and shook his head. "No."

Cal stared at Malcolm for an uncomfortable period of time and looked back at the windows. "Must have made a bit of noise."

"My bedroom is in back."

"You were up and about the whole time?"

"No, only the last two shots."

"Good thing no one got hurt." Cal pivoted to gain sight of the road. "Mind if I take a look inside?"

"Not at all, come on in." Malcolm led the way through the door.

Precious greeted the visitor with a wagging tail. "Cup of coffee, Chief?"

"Oh, no thanks." Malcolm guided Cal to the guest room. The chief surveyed the space and frowned. Rolling the temple of his sunglasses between his fingers, Cal approached the window and surveyed the walls and floor. "How about the other room?"

"Next room down the hallway."

Malcolm remained in the doorway as Cal ambled to the windows. "Two shots in this room also, Doc?"

"I think so." Precious sauntered in and sat beneath the windows, her eyes tracking the chief.

Cal placed his hands on his hips as he redirected his interest to the interior walls. "This is your office?"

"Yes." Malcolm rocked on the balls of his feet.

Cal warbled a two-toned whistle and scratched the drywall opposite the window. Reaching into a jacket pocket, he produced a pencil and jabbed it into a small hole. "Doc, I would guess that one of the bullets sailed this way. No stud here, so the slug probably passed through the inside wall as well."

Malcolm asked, "You mean ricocheting around the house?"

Cal scrutinized the divot. "Makes you think a little bit, doesn't it?"

Malcolm exited through the front door with Cal and halted while the chief meandered across the front lawn, eyes trained on the ground. Reaching the lawn's edge, he waited to cross the pavement as a pickup sped by on the road. Cal stopped to stir the loose gravel in the far shoulder. His casual pace and detached attitude upon returning to the house was starting to rile Malcolm.

Cal opened a clenched fist. "How about that, Doc? A couple of .22 shells." The shiny casings looked benign in the palm of his hand. Cal shoved the trinkets into his pants pocket and turned back to his car.

"Chief, how did you know to come here?"

Cal leaned heavily on his door. "You've managed to get yourself in a bit of a dust up with some folks around here."

Malcolm was tiring of the detective game. "What folks?"

Cal said in a reproaching tone, "Got a fair-sized list of grievances against you for a newcomer."

"From who?"

"Let's not play games, Doc. For starters, I heard through the grapevine about your visit with the dog. And that prank went and got a little out of hand didn't it? You didn't heed my advice earlier and now things have escalated to where someone is going to get hurt…and hurt badly."

Malcolm glanced back through the storm door at Precious sitting on the inside. "What is going on, Chief?"

"Doc, I realize you're new around here. But you're keeping company with a group of local instigators. I suggest you avoid them." Cal returned to his car but halted before getting in. "And for God's sake, mind your own business. Some folks in these parts don't take kindly to people who stray into theirs. Have a good day." Cal slid behind the wheel and slammed the door. The gesture, Malcolm thought, had an air of judgement and finality. The Ford backed off the drive and descended into town.

An hour later, Malcolm laid the final damaged windowpane in the back seat of Lucille. Precious rested in the passenger seat with a soup bone between her paws, a treat from a recent trip to the butcher shop.

"I thought you buried that damn thing out in the yard." Precious sat up and wagged her tail. "Let's go to town, bonehead." The short ride ended in front of a rambling two story structure embellished with a faded burgundy and white-striped awning. A sign above the awning, which clung to the wall forty-five degrees off horizontal because of missing screws, read 'Carling's Merchandise and Farm Supply.' Tethered to a rail, a pair of horses loafed with heads lowered and one rear hoof knuckled off the ground, enjoying a temporary respite from pulling the white top buggies behind them.

Malcolm retrieved the glass while Precious sprung out of Lucille and waited by the entrance. She was flanked by canning jars, yard rakes, chain saws, axes, and orange hunter vests that decorated the limited space inside the alcove windows. Bracing the glass in one hand like a stack of books, Malcolm gently pulled the screen door open, which slammed behind him as he creaked across the rough wooden floor. A teenager with red hair streaked with purple was engrossed in a Cosmopolitan magazine, sitting behind the counter. Her head was bobbing to music from a Walkman that was strapped to her ears. Behind her, an old St. Bernard snoozed on a braided rug.

Malcolm laid the glass onto the counter. "Hi, can you help me?" The teen didn't lift her eyes from her reading. Malcolm waved his hand in a circle to gain her attention.

The counter keeper looked at him with disdain and pulled off her ear phones. "Did you need something?"

"How about some replacement glass for windows?"

She examined the collection closely and rolled her eyes while popping a big bubble of her gum. "I can't fix this."

"Where's Mom and Dad?"

"I don't know, gone out for a while."

"Could you please see that they cut some replacements when they return?"

"I guess." She went to reattach her earphones.

Malcolm held up his hand. "In the meantime, how about four yards of plastic sheeting from a three-foot roll?"

With a pronounced sigh, the girl stood up and looking as if she was asked to clean unflushed toilets, sauntered off to the back room.

Malcolm offered in an polite voice, "Thanks so much, I knew you were the right person for the job."

The slumbering canine giant lifted his head to make note of her departure. Malcolm reached a hand into a glass jar on the counter and fished out a dog biscuit, which Precious snarfed out of mid-air.

"Woof!"

"That's enough! I'm not Lois."

A couple of Amish women in calf-length blue dresses and straw bonnets with scarves were sizing up the canine antics over the top rim of their eyeglasses. An awkward smile from Malcolm persuaded them to return their interest to canning supplies. Jonas P. Yoder strolled past a display of dowel rods, holding a brown paper bag of nails that he set on a metal pan scale. "Well, haloo veterinary."

"Hello Jonas, how are things at your place?"

"Well enough, I guess."

"Sheep doing well?"

Jonas P. seemed surprised. "You must have a little of the hexing in you."

"How so?"

"Well, one hasn't been eating right the last few days."

Malcolm wrinkled his face. "Any others in the flock looking sick too?"

"No, not that I can see."

The girl reappeared with the plastic, poorly folded. Malcolm paid for the sheeting, bid his farewells and left the store, Precious at his heels. Back in the Scout, Precious reacquainted herself with the soup bone. Malcolm turned the key, shifted Lucille into reverse, but kept his foot on the brake, engine idling. Jonas P. came out of the store shortly. Malcolm pushed the gear shift back into park, stepped out of the door, and called, "Pardon me Jonas, you said that all your other ewes are doing okay?"

Jonas P. halted, pondered. "Well now that you mention it, one died a while ago."

"Died?" Malcolm's voice raised half an octave.

"Ja, about a month back."

"Have any idea why?"

"Didn't want to eat much." Jonas P. chuckled, "It was acting like the ram, chasing the others around the pen all the time."

Malcolm slid back into his seat. Tapping his middle finger on the steering wheel, he watched Jonas P. untie the reins and climb into the buggy. Precious placidly gnawed on the bone. "Sometimes I wish my life was as simple as yours." He put the Scout into reverse, "Time to staple plastic before our company arrives."

CHAPTER XXVI

Malcolm nursed a mug of coffee on his patio, watching Precious tracking scents about the the yard. Malcolm sniffed to prevent his nose from running. Better pop an antihistamine before I go. The first hard freeze of the season would be a welcome event.

A hoary blur ripped across the side yard on the far side of the fence and circled back out of sight. A terse female voice yelled, "Pierre! Pierre, come back here *now*!" Precious made a bee line for the fence, thrusting her nose between the slats. Malcolm finished his coffee. "Come on inside sweetie pie, time to meet a new friend."

Precious whined by the front door, waiting for Malcolm. He opened the door and took a deep breath. Jolene had a clamp on Pierre's collar, restraining him on the side of the road. Malcolm said, "Stay here for a second." Precious barked and pranced as he left her in the house. "Need a leash?"

Jolene gained ground away from the roadside, yanking the sheepdog with a taut forearm. "He doesn't take to leashes well."

Malcolm surmised. I'd never have guessed. "I didn't want him getting away from you, traffic rolls by at a good clip on this road."

"He won't." Jolene prevailed, her auburn mane kept in check by a ponytail. "He can be a bit of an ass when he gets excited. He usually has the whole farm to release his energy."

Malcolm opened the front door for his guests. Pierre roared through the threshold to introduce himself to Precious, who reciprocated by growling with an icy glare. Pierre balked and retreated around the furniture. "He just wants to play," Jolene said. Gaining momentum, the grey and white meteor galloped towards Malcolm. With the avidity of a mongoose, Precious snarled and barred the bigger dog's path. Pierre stopped in his tracks and retreated with his head down.

"She's got an attitude."

Malcolm stroked Precious' head. "I think they've worked out who's the alpha." He poured coffee into a mug and handed it her.

She opened the refrigerator and retrieved a carton of milk. "Embracing Galloway County lifestyle with plastic sheeting over the windows for winter?"

"With a matching ensemble of bullet holes."

She splashed milk into her coffee without taking her eyes off his. "Congratulations, you're becoming more like a local every day. Pop a gun rack into your Scout, start chewing tobacco, and you'll be all set."

"Apparently target practice on people's homes is common around here?"

"School windows, mostly BB guns." Her eyes wandered towards the ceiling, "Tool sheds at construction sites, road signs, especially the deer crossing signs." She set the milk carton onto the counter. "You sure someone targeted you specifically?"

"Chief Cal seemed to think so."

"Police work already? When did you call him?"

"I didn't, he just showed up. Kind of mysterious, asking me questions as if I knew why it happened."

She sipped her java. "Well, do you?"

Malcolm recounted the tale of his evening at MacDougall's homestead, finishing with the observation, "That was several weeks ago."

"It's been said revenge is a dish best served cold." She swirled the coffee in her mug. "You really didn't see anyone?"

"I was asleep, Precious woke me up."

"Asleep?"

"My room is in the back of the house."

"So Chief Cal didn't offer any information?"

"No, seemed a little perturbed with me."

"Wonder why?"

"All I wanted to do was examine that damn black dog, getting in the middle of a family feud was the last thing on my mind."

Jolene bent over to replace the milk in the refrigerator. Malcolm watched the flex of her hips.

"Malcolm, don't kid yourself, up among these hills, people hold grudges for a long time." She closed the refrigerator door and rotated to face him. "A *very* long time."

"I was just trying to do the right thing."

She shook her head. "The right thing was not to go onto somebody's property for the benefit of the Calhoun boys."

"I needed their help."

Her hand reached out and held his. "The hostilities between the MacDougall and Calhoun clans started long ago from squabbles over land and rolled right into the bootlegging days. Folks like that are always willing to drag in new recruits, even if they're put in harm's way. You might say it's a sporting proposition to them. That sortie to MacDougall's was an initiation into the Calhoun club."

"I don't think there was any devious intent."

Her hand tightened to a squeeze. "This isn't a prank like scampering away after ringing doorbells. Around here, it's hard to tell participation from manipulation. And when ignorance and suspicion are involved, that leads to violence."

"I've known people like the MacDougalls everywhere."

"But Galloway County is different, Malcolm. Change in this neck of the woods comes slow, and more often than not, passes on by."

Malcolm regarded the growing warmth of the early afternoon sun outside. "It's a beautiful day. I thought we could take the Scout. This way, our furry friends can sit in the back seat if they get muddy."

"Would your vehicle pass state inspection?"

"Present condition, maybe not."

She peered over the top of her mug, "Okay, I'm kind of a risk taker."

They placed their dishes in the sink and made for Lucille.

Precious claimed the middle of the rear seat. Jolene corralled Pierre next to his new playmate. He stretched his neck to lick Precious on the snout. She curled her lip and growled. Pierre backed off, venting his frustration by barking.

Jolene spilled into the passenger seat. Malcolm got in last, found the keys still hanging in the ignition and cranked the engine. He backed down the driveway and cut the wheel onto the road. Lucille's horn blared, ceasing only when Malcolm set the gear to drive and straightened the wheel.

"What's with that?"

Bewildered, he rubbed his neck with one hand. "That's a new feature. I had her in for service this past week to fix the ignition."

"Kind of quirky, is she?"

He shrugged. "Never know with old cars."

They drove through Salty Dog, where an ample number of vehicles were already filling the parking lot of the Dead Bear. Leaving town, they trekked along the highway that twined between two ridges on the northern edge of the valley. Jolene directed Malcolm along a dirt road that cut through the floor of a narrow glen, paralleled by a lively stream. Above the track, a kaleidoscope of color and shadow sifted through a lofty canopy. Rhododendron flourished underneath, glossy green leaves adorning their twisted limbs. After ascending for a mile, the road emptied into a large clearing next to a tranquil pond constrained by a rock dam. For a moment, they paused in Lucille to indulge themselves with the reflection of trees on the placid water.

Malcolm gave in to pent-up canine anxiety and the dogs burst out of the rear door. He raised one eyebrow. She replied, "He's pretty good about staying in earshot. Just needs to release his steam." Pierre tore straight for the pond, disrupting the mirrored still life with a headlong plunge.

Malcolm said, "Like I said, our companions might get a little muddy." He shouldered a knapsack, leaving Lucille under the shade of a maple. The scrunch of gravel under their shoes changed to a soft thump on brick-red soil as they veered onto a trail that threaded through knee-high grass. Precious fell into pace a few feet ahead. Pierre darted by the entire party and disappeared up the trail. Malcolm thought dogs were like small children in so many ways.

Calf muscles strained as the trail ascended the ridge overlooking the pond. The spell of vibrating leaves and dappled sunlight remained unbroken by conversation. Time and distance were marked by the occasional yellow leaf floating onto the path. Precious maintained her position as advance guard, Pierre sporadically returned at a canter, only to follow another tangent into the woods.

Malcolm asked, "What's with the clearing and old buildings back by the pond?"

"This was an old CCC base camp from back in the Great Depression. My uncle worked in one as a young man. They built the roads, picnic areas, and ranger stations up in these mountains. There are signs all around us of their work, including the trail you're hiking on.

"It seems untouched, fifty years later."

"If you close your eyes and touch the stone walls they built, you can still feel their sweat and aching muscles."

"You believe in ghosts?"

"In certain places, yes. More of a presence than a ghost. And you?"

An awareness of not wanting to be deep within the forested ridges after dark wriggled down Malcolm's back. "Perhaps in old places near old graves."

Up ahead, Pierre ran across the path from one side of the undergrowth to the other. They halted at a ninety-degree turn in the trail. From this point, the path gradually climbed along the side of the ridge, rather than by direct confrontation. Malcolm stepped one foot onto a small boulder and rested a forearm on the raised knee. He tried to reckon the elevation from the camp below.

She asked, "How are things going with the rabies?"

He hesitated. "No more cases that I know of, yet."

"I hope that doesn't include the MacDougall's, that place has a bad reputation in the valley and should be avoided."

"I kind of found that out the hard way."

"That family has a stranger history than most."

"Now we're getting to the real ghost stories."

"These mountains and valleys have seen their share of hardship and misery. It's difficult for an outsider to understand."

"I like your friendly CCC spirits better."

She smiled. "There's a good spot to rest a little farther up..."

The path leveled off to an even grade. To their left, four hundred feet below them, lay the camp within the glen. A large flat rock, dotted with plaques of lichen, rested on the lip of the ledge. "Nice place to hang out for a while," she said. Malcolm wavered as he looked over the edge. She held his hand. "You afraid of heights?"

"A little, although thrill rides at amusement parks are worse."

"I love roller coasters."

"And from a farm perspective, silos don't excite me either. Henry tells me that one of his clients was killed after falling from one the year before I came to the valley. Didn't you tell me that you ended up back here because of a similar episode with your dad?"

"That was one of the reasons."

"It isn't only the height of silos, as much as the confined space inside."

"Claustrophobic, as well?"

"Don't mind elevators. But wouldn't ever be a spelunker."

Her eyes flared. "Lots of caves around here, use to explore some of them as a kid. My brothers and I use to make a mess of the stacked hay bales in my dad's barn. We'd make tunnels with them and spend hours crawling around in the dark."

"Like rats in sewers?"

"I wouldn't exactly use that analogy."

"Groundhogs?"

She sank down on the stone shelf, hanging her feet over the lip. Easing beside her, he produced grapes, cheese, Lebanon bologna, and bread from the pack. Precious instantly found a place next to him. "She's such a beggar," he said as a morsel of cheese disappeared from his palm. Precious shifted her eyes from one human to the other in a culinary vigil. In contrast, Pierre wandered by on an irregular basis, happy to continue his exploration of the brush.

Jolene chewed on her sandwich and her thoughts. "This county has an inherently distrustful mindset."

"About what?"

"Non-conformity." She took a long look at him. "There are limits to a cow doctor's role in local drama you know."

"Sorting out the location of rabid animals is well within those limits."

"But you can't go around acting like the police. If you get in trouble you'll have yourself to thank for it." Malcolm dwelled on her words. She asked, "You told me last week at dinner that you got divorced, why?"

The change in direction caught him off guard. A cherry leaf fluttered onto his lap. He pinched the leaf between his fingers and examined the pattern of brown spots against a background of faded yellow. "Clash of cultures more than anything else, I guess." He sipped some water. "Her father was a big time accountant in Chicago. Family lived in the wealthy, north shore suburbs."

"How did your worlds come together?"

"College. We shared a class about rapacious Greek gods and sculptures of naked warriors."

"You mean mythology?"

"I was always fascinated with those stories. Towards the end of the semester, I met her at a party."

"Seems to be your style."

"Yeah, well. Her folks didn't care for me much. I was a little too off beat."

"This sounds vaguely familiar."

"I think she found me refreshing…at first."

"Refreshing or sophomoric?"

"Hey, back then, everyone was questioning authority. I thought she wanted to join the parade. But then, there were those all-to-frequent visits to her parent's social functions, where insincere people size up each other's cars, vacation properties, and even spouses."

"The way of the world."

Malcolm nodded. "Yeah, I'm afraid so. I'm not a malcontent, Jolene. But in *that* world, the clothes that are worn at the gala events get more attention than the charity."

"What's her name?"

"Carrie."

"And she wanted to return to that world?"

"I think college was a little retreat to see how the commoners live. The *coup de grace* was my job out here after vet school. I wanted to get away from her parents, and mine, too, I guess. I was a bit guilty of a 'follow my career or don't follow at all' ultimatum."

Her eyes widened. "Yeah, kind of…"

Malcolm nibbled distractedly on his sandwich. Tired of eating, he donated the remainder to Precious. "In the end, she decided that the male role model in her life should be more like her dad rather than me. She's following her chosen path. I wish her well."

"There were no kids?"

"Oh God, no! We both knew our relationship was unsettled."

"Did you love her?"

"Head over heels."

"Do you still?"

He sucked in a lungful of air. "I thought I did, but now I'm not so sure."

She straightened her back and looked away from him across the valley. He laid his hand on hers. "But it's over. I even have the lawyer bills to prove it."

She turned; her eyes scanned the contours of his face for veracity. Precious lay snoozing in the sun, dreaming of cheese and bologna. "Have you been with anyone since?"

"No, too many hours on farms and in the car." He laid back on the rock and watched a lone cloud mosey past the tops of the trees.

Her hand hovered over his chest, skimming the surface with

each ebb and flow of breath. She raised her palm; leaving only her fingertips on his torso. She brushed his shoulders and neck, followed by a profile of his cheeks and brow. Softer than worn silk, a finger brushed his lips. She leaned on top of him and kissed his forehead. Malcolm ascended into the warm autumn sunshine.

*****

Back home, drowsy dogs lay on the floor, worn out by the hike and fresh air. Jolene was taken aback by a hefty orange tomcat that sniffed at the canine louts, sauntered into the kitchen, and progressed to a water dish. The tabby strutted with pride, tattered ears marked his many battles.

"Who is that?"

Malcolm waved a hand ceremoniously. "Jolene, may I introduce you to Bonnie Prince Charlie, the true contender for the throne of Scotland."

"He looks rather comfortable with his surroundings. It doesn't look like he misses many meals, either."

"Few if any, I should think."

"How old is he?"

"Don't know."

"You got him as a stray?"

"Not my cat."

"Not yours?"

"Neighbor lady's. Miss Violet. She keeps company with Precious on days when I can't drop her off in the office. Miss Violet watches her soap operas here; doesn't subscribe to cable TV and she can't watch her favorite shows on the two stations we get over the air. Prince Charlie is usually Miss Violet's calling card, along with the occasional casserole in my refrigerator."

"How long does he stay?"

Malcolm bent over to stroke the cat. "As long as he sees fit, usually, a day at the most."

Bonnie Prince Charlie assessed the sorry plight of the human

155

peasantry and exited with the same disinterest as when he entered.

"Kind of odd to come over and use your TV."

"She's elderly. It's not unlike the Amish using phones in someone's barn or house to make a call. Besides, some farm calls aren't conducive for me to bring Precious and I won't take her on cold, or especially hot days to sit in a car."

Jolene leaned against the wall, arms folded, legs crossed; the epitome of stillness save for her forearms levitating with each breath. Pierre and Precious snored blissfully on the floor, the only audible signs of life in the house.

Malcolm ran his fingers through his hair. "I could throw some steaks on the grill and open a bottle of wine."

"I suspect what you really would like is to continue where we left off up in the woods?" She laid her arms on top of his shoulders and pressed her hips into his.

The hardwood floor of the hallway to the bedroom was littered with discarded shoes and garments. Jolene leaned back on her elbows across the bed, the pastel tones of the setting sun shading her alabaster torso. Malcolm knelt astride her abdomen as she stretched her arms above her head and grabbed the headboard. Coiling her legs, she wrapped them around the small of his back, emerald eyes igniting within a halo of disarrayed hair as he entered her. She clamped him ever tighter, not letting go until their passions were spent.

As Malcolm lay sleeping on his side, Jolene watched the last of the twilight settle into night through the window. There was something about the MacDougalls and Calhouns that needed to be flushed out from her memory. She would research some of the old records in the newspaper office. Malcolm might be in over his head more than he knew, and presently more than she knew as well. Listening to his measured breathing, she realized that her feelings for him were growing deeper, and yet she wondered; was he truly resistant to the influence of the valley culture, or would it change him in the end, like it had so many others.

CHAPTER XXVII

Jonas P. Yoder balanced on a stool, illuminated by the flicker of a kerosene lantern that could have been a relic from the days of steam locomotion. A splish-splash of milk drummed a metal pail between his legs. The farmer broke his concentration to heed Fritz as the dog bolted out the stable door and into the crisp autumn air. A vehicle had parked in the barn yard.

Malcolm extended his feet from behind the wheel; another long day was at an end. His reliable supply of brown eggs would soon dwindle with the waning daylight. One of Jonas P.'s border collies greeted him with a wagging tail, sniffed his boots and lifted a leg on Lucille's tires before trotting back to the barn. Good thing Miss Violet was babysitting Precious today. He heard footsteps on the wooden porch of the farmhouse as Rachel came down the steps and onto the walk, her eyes bright and alert.

"A dozen eggs today if you have them please, Rachel. Got any produce left from the frost?"

The girl swung open the yard gate, eyelashes blinking behind her glasses. "Some squash and onions."

"Great, I'll take a couple of each, please."

A shy smile lifted across her lips as she entered the milk house. She reappeared with a brown paper bag. Malcolm handed her three bucks. "Hope your hens keep up the good work, fresh eggs get tougher to lay this time of year."

"Thank-you," she said and returned towards the house.

Malcolm called out, *"Guten abend."*

The gate hinges creaked as she turned and smiled, then continued on her way.

Malcolm nestled his produce collection in Lucille and homed in on the light of the stable. As he entered the stuffy byre, Jonas P.'s sons peered at the visitor around the points of soiled rear legs.

"Ah, *Tierarzt, wie bist du?*"

"Well enough Jonas, and you?"

"Good." Jonas P.'s calloused hands pumped a pair of teats like pistons. While milk from the first spigot splashed into the pail, the next stream was already in mid-air. Malcolm asked, "How is the ewe you told me about at the hardware store a while back?"

Jonas P. extracted himself from under the cow and stretched his back, pail of milk in hand. "About the same. But it's not a ewe, it's the ram."

"Oh, where is he?"

"In a pen out in the shed."

"By himself?"

"*Ja.*"

"Why is that?"

"He's wanting to be a little *too* rammy with the ewes and spring lambs; head butts them if they get too close."

Malcolm creased his brow. "Mind if I take a look?"

"Suit yourself. In a little bit, I'll be out." Malcolm nodded and retrieved a flashlight. A cloudbank was prematurely swallowing the sun on the horizon, turning light into shadow.

The flock scurried about in a fluid wave of wool as Malcolm entered the paddock. Malcolm targeted a choppy bleat echoing from within the shed. At a glance, the ram seemed clinically normal, perhaps a little thin in the flanks. Nervous from his isolation, the ram bobbed his head and hastened to the far corner, staring at the unfamiliar human, snorting through his nostrils. Malcolm reasoned to wait until Jonas P. was done milking before trying a physical exam. A chill breeze touched Malcolm's cheek as the sun departed and painted a saffron aurora about the ridges. The farmer's boots tamped the earth behind Malcolm. "Shall we take a look at him, veterinary?"

The duo climbed over the fence and directed the sheep into a corner of the pen. Malcolm pushed the halter over his head. The ram tugged against the rope, swinging his head and yanking Malcolm into a side wall. Malcolm managed to wrap the loose end around a post and handed the rope to Jonas P.

Finished with his exam and shoving his stethoscope into a side pocket of his coveralls, Malcolm frowned. "Not much that I can tell, Jonas." The ram continued to struggle against the halter.

Malcolm released his patient. The ram wrestled out of the restraint, backed away, stamped his forefeet and bobbed his head. Jonas P. chuckled, "You better look out veterinary, looks like he has a mind to chase you out of the pen." On a whim, Malcolm stepped towards the ram. The sheep stood his ground and stamped his feet. "That is strange,"Jonas P. mumbled. "Holding his ground like that."

Malcolm advanced another pace; the ram inched back but didn't take flight. Malcolm's chest tightened in a slow, cold squeeze. "How long has he been a little off Jonas?"

"Oh, a couple, maybe three weeks."

"You mentioned the other ewe that died, after the dog attack?"

Jonas P. registered the events in his mind for a few moments. "*Ja*, maybe three weeks ago, more than two months after that day."

"All the other sheep doing okay?"

The farmer surveyed his flock on the opposite side of the paddock. "I think so."

Malcolm felt a foreboding as he looked at his hand which held the halter. "Jonas, have you been handling him much?"

"No."

"I would like you to keep away, feed and water him over the fence. What does he eat?"

"A little hay, maybe."

Malcolm studied Jonas P. for a long count. The farmer reached into a pocket and pulled out a sack of tobacco to roll a cigarette. Malcolm clambered out of the shed pen. "He's acting a little like an animal with rabies. I'm going to stop by in a day or so, to make sure I'm not jumping to conclusions."

Jonas P. lit his smoke. "Okay, *Tierarzt*, but when do you think you'll know?"

Malcolm replied, "I imagine by the time I visit again."

In the milk house, Malcolm scrubbed the halter and his hands in betadine as if he were preparing for open heart surgery. He then scrubbed the halter.

Jonas P. was waiting by Lucille. "I see you got your eggs from Rachel."

"She's good help."

The farmer imperceptibly stood a little taller. "She likes to read books. She started to work this fall as Sarah Hostetler's helper."

"Sarah Hostetler?"

"Teacher at our school. She will retire in a couple of years. Rachel is going to be the new teacher."

"You must be proud of her."

"Ja, she teaches her younger brothers and sisters to read good. *Und jetzt*, the school children."

Malcolm settled into Lucille. "Take care and I'll be back soon to look in on the ram, Jonas."

Jonas P. raised his hand, "*Guten abend, Tierarzt.*"

Malcolm halted at the end of the farm lane. He turned to look up the road in the direction of the MacDougall place and wondered about the black dog. Turning towards home he encountered a pair of headlights coming his way. As the oncoming vehicle got nearer the high-beam lamps came on, forcing Malcolm to squint and look away. With a burst of speed, the approaching vehicle swerved across the center line, only to wheel back into the left lane within feet of Lucille. A white pickup flashed by, and a gruff voice yelled, "SHEEP FUCKER!" Malcolm hit the brakes and watched the taillights of his assailant shrink in his rearview mirror, past Jonas P.'s lane and on up the road. Malcolm pitied Jonas P. and his proximity to his bad-tempered neighbor. It was more than keeping the faith; it was a test of maintaining sanity.

CHAPTER XXVIII

Nestled on a bench between shelves of drugs, Henry followed Malcolm's pacing like a tennis fan glued to a highly contested volley. His young associate was trying to persuade Henry, Lois, and the retired founder of the practice, Paul Kramer, that the rabies outbreak in the valley was following a disturbing path. Lois sat at the desk, Paul standing behind her. Malcolm tried to connect the attack on Jonas P.'s sheep, some Amish women in a buggy, and now a sick ram to John MacDougall's dog. Henry noted that Malcolm didn't include the visit to MacDougall's homestead, an incident that Malcolm thought had remained unknown to his employer. *There are no secrets with Lois, Malcolm.* The panel allowed Malcolm to finish his testimony before comment.

Henry, finishing his warmed doughnut, was the first to speak. "Perhaps the attack on Jonas P.'s sheep has developed more problems than originally thought."

"Rabies could be a common link between all of this," Malcolm said. "Something isn't adding up. My way of thinking from school and from John Y.'s cow, is that once clinical signs begin, the disease progresses rapidly."

"The virus has been known to have long incubation periods in some species," Henry said.

"Yes, but shedding is supposed to be within a relatively short time before the end, at least in dogs. The sheep didn't die until a couple of months after the attack by MacDougall's dog."

Paul folded his arms. "Well, this thought of sheep dying from rabies months after an attack by a *presumed* rabid dog borders on science fiction. The only confirmed *case* of rabies in this valley was

John Y.'s cow. She progressed in a few days, just like the textbooks say it should. What makes you think this black dog is any more aggressive than normal? After all, he chased you into a tree. And may I add that's a normal response to a trespasser who didn't belong there in the first place."

Malcolm's face reddened. "It's an impression that Dale and Sally Calhoun, and some of the folks in the Dead Bear..."

Paul scolded, "Since when are a bunch of barflies considered to be an acceptable source of a professional, scientifically based diagnosis?"

Henry intervened. "Jonas P.'s ram has been off feed for more than a week?"

"At least two."

A surge of fresh air invigorated the room with the arrival of a bearded figure in a straw hat and black denim coat, Daniel L. Peachey. "Good morning, veterinaries." A collective salutation greeted the visitor. "I need some horse wormer and white lotion for the legs."

"How many horses, Dan?" Malcolm asked, receding into the back room.

"Five Belgians, two driving horses, and a yearling."

Malcolm grabbed an armful of worm paste tubes and a pint bottle of a chalky suspension. He deposited them on the desk, as Lois wrote up an invoice. Fetching a brown paper bag, Malcolm asked, "Anything else?"

"No, that's plenty from the pocketbook today," Daniel L. smirked.

Henry queried, "Are you done picking corn, Dan?"

"Almost, Mr. Breck, another five acres or so."

"No weddings yesterday?" Lois inquired as she handed Dan L. the invoice.

Dan L. looked at the slip, pulled out a hundred-dollar bill, handed it to Lois. "No, not in our group. We have one tomorrow, John S. Peachey and Sarah Byler."

"That would be Steven M. Peachey's son and Raymond R.

Byler's daughter?"

Daniel L. laughed. "You keep up on the goings on in the valley."

Lois handed Dan L. his change, "We try as best we can," and glimpsed at Malcolm. Daniel L. wished them a good day and exited the office.

"I would like to check up on that ram, at least every other day or so," Malcolm said. "I'll make it my personal project. Jonas P. lives out my way, anyhow."

"There are other chronic neurologic diseases in sheep," Henry reminded him.

"I'll keep an open mind."

"Let's not have your lack of judgment conjure up hysteria in the valley." Paul added curtly.

Malcolm bit his lower lip and walked out the door.

Paul thumped his fist on the desk. "I told you that he's stirring up trouble. That divorce business within a year of coming to the valley and now carrying on with the drunken deadbeats in Salty Dog."

Lois chided, "Now Paul, he is a little green, but he has a good heart."

"He's supposed to be a professional. People will talk."

Henry pushed up and away from the bench, drained his coffee. "Paul, he has a good head on his shoulders and is a hard worker. Sometimes, I think a little too hard."

"He's headstrong and has a different attitude about life."

"As was I when you hired me twenty years ago?"

"I'd tell him to stop these shenanigans or find another job."

Henry gazed at Paul. "But this is my call to make now."

Lois said, "What if that young man is on to something about this rabies stuff? Have you thought of that?"

Henry's eyes slid towards the door. "Yeah, I have."

Lois weighed Malcolm's willingness to bolt into circumstances he didn't fully understand. Jolene Brubaker came to mind.

*****

The morning sunshine gradually weakened into a filmy veil. An east wind agitated purple and blue dresses, white shirts, and black pants on the laundry lines in farm yards.

The air bit at Malcolm's nose as he judged the clinical status of his patient. The ram was the same as the other day, suspicious of Malcolm and stamping his feet. "Well, if you had rabies like John Y.'s cow, you'd be dead by now." Shaking his head, Malcolm returned to the stable.

Jonas P. was pitching hay in front of his hungry herd. Leaning on the wooden handle of his fork, the farmer waited for an opinion. Malcolm said, "Nothing to suggest a change since I was last here. I'll be back in two or three days."

"*Sehr gut*, veterinary."

Before leaving, Malcolm placed a half dozen elementary reading books on top of the cooler in the milk house, a small treasure he had discovered at the library book sale the past weekend. He penned a note. '*For Rachel, you will make a fine teacher.*' He turned to Precious as he cranked Lucille's ignition. "Hope they teach reading in English as well as German."

A message awaited Malcolm on his answering machine. What did he think of having dinner at Jolene's folk's place this Sunday? He called Jolene, she suggested a rendezvous at her apartment. "Relax Malcolm, you won't have to provide the entertainment, my folks aren't the type to socialize over cocktails for an hour before eating, we just show up and eat."

Later that night, Malcolm stretched his feet under the covers, his hands clasped under his head. A cold rain strummed the windows. Precious snored, lounging on a fleece rug next to the bed. What would dinner with Delmar be like? Certainly, different from dinner with Carrie's folks; mind traps that tested both his knowledge of elaborate sequences of dining utensils and patience with judgmental diatribes. If nothing else, the farm menu would likely offer a more simple fare—and better digestion.

CHAPTER XXIX

The mountains had removed their verdant summer wardrobe in exchange for the drab attire of winter. Shadows slid across their bronzed torsos, underneath a mix of sun and roving clouds. With each swell of the breeze, arboreal fingertips shed flecks of color onto the road.

"So, what does Delmar think of you dating the vet?"

"You mean is Dad impressed with your professional degree?"

"I mean, does he feel a little uncomfortable with someone who comes to the farm on business and dates his daughter?"

She watched a leaf spin in a turbulent arc across the wind shield. "Dad thinks you're okay, but don't let it go to your head."

Malcolm slowed down to pull Lucille in front of a two-story farm house built of solid burnt-orange brick. Jolene popped out of Lucille and up the steps of the porch. Pierre celebrated her arrival by cantering to one end of the porch and cannon-balling back towards the visitors. Malcolm raised his leg to offer a salutatory knee to his well-wisher. Rocking from the collision, Malcolm grunted, "Same old lively self." Pierre hopped onto his hind feet. Malcolm boxed against two bushy paws while the dog joined him in an awkward dance of conflicting desires for intimacy.

"Are you done fooling around with the dog?" Jolene asked as she opened the door and the warm, moist scent of comfort food spilled outwards. Passing through the foyer, they entered into the living room. A couple of overstuffed chairs and a couch, all of dissimilar fabric, occupied much of the faded hardwood floor. A wood stove urged the guests to shed coats and hats and bask in comfort. Large magenta and ivory magnolias, cloaked in shiny green foliage, gushed from the wallpaper. The Brubaker home was a house of living. Jolene called out to announce their arrival. Malcolm followed her around the large oak dining table and into the kitchen.

Delmar was sequestered among magazines and newspapers littered about the table. Jolene breezed over to him and gave him a hug. "Well, Doc, you got to join us after all," Delmar beamed.

"Hello, Delmar." Malcolm took notice of the woman by the stove, preoccupied with stirring thick brown gravy. Her striking features bore evidence of her daughter's genetics. Despite a life of farming, her green eyes had the same incisive gaze from within a mane of auburn hair. Her posture was erect, her movement balanced. Malcolm bowed. "Mrs. Brubaker, how do you do?"

Her lips spread into a warm smile. "Millie would be preferred over 'Mrs. Brubaker', Malcolm."

"Yes, ma'am, Millie it is." Jolene, positioned behind her mother, mimicked 'yes ma'am'. Oblivious of her daughter's antics, Millie targeted Malcolm with her spoon. "I hope you brought an appetite."

"Yes, ma'am, I did. It smells delicious."

Jolene sighed and rolled her eyes.

"Doc, I was just reading to the missus that all those space satellites and shuttles are affecting the global weather."

"I haven't heard that, Delmar."

"Well, I've got that article around here somewhere. They're shooting all sorts of things up there, TV and weather satellites, rockets with thermal rays and laser beams, run by specially trained robots. Damn government's got their eyes everywhere. They can read the license plate on your Scout, Doc." Delmar sifted through the pile on the table.

"I didn't know I was of that much of interest to them."

"Delmar, be a dear and open up a bottle of wine will you?" Millie interposed.

"I *know* it's here," he mumbled, dismayed by the edict to suspend his foraging. The farmer raised himself stiffly. "They say that this space junk reflects the sun's energy away from Earth.

There's so many of them, not as much light gets through the starmosphere." Malcolm, unsure how to respond to Delmar's unique terminology, did wonder however, as to the quantity of man-made debris floating about in orbit…and when it might fall back down.

"Jolene, would you mind getting the table ready?" Millie asked.

"I'll help", volunteered Malcolm. Lugging a stack of plates hoisted upon him by Jolene, he tramped to the dining room, the old floorboards creaking under his shoes. Jolene followed with silverware and directed five settings. Malcolm looked at her.

"What, no 'yes ma'am' for me?" Soon after, they volleyed bowls and pans with mounds of mashed potatoes, green beans, applesauce, coleslaw, corn, rolls, and brown gravy thick enough to walk on. Delmar sauntered into the room with a bottle of wine and glasses, still chatting about the flow of energy in the cosmos. Millie appeared with a roast beef, cooked well done. A slight, wiry youth walked in, Jolene's brother Marty. Malcolm recognized him from the barn, nodded hello.

Delmar sat at the ceremonial seat of honor, grasping the carving set by the roast platter. As heads bowed, Delmar beseeched, "Lord, thank you for the food which is spread before us, bless it for our use, and us to thy service, Amen."

Millie poured herself some wine. "Jolene tells me you're originally from Illinois, Malcolm."

"Yes, ma'am, outside Chicago."

She savored a sip from her glass. "Delmar and I took a trip out there soon after we were married. Much to see in that town."

Delmar added, "Lots of trains, Doc. Me and my brother, Chub, hopped freight trains all the way out to Denver one summer, living off the land and working odd jobs. Chub did a little motorcycle racing. Found ourselves an old Indian Chief in a scrap yard and got it running."

Delmar parried a tidbit of roast on his fork in Malcolm's direction. "You ought to do that sometime, Doc. Take a train to see the country. Of course, nowadays, you couldn't be hoppin' trains the way we did."

Malcolm admitted, "As a kid, I always wanted to be a railroad engineer."

Millie laughed. "That's a couple steps removed from a large animal vet."

Delmar recalled his western geography, often scrambling the proximity of the Mississippi River relative to state borders. Jolene remained uncommonly quiet, content to creep her foot against the inside of Malcolm's leg whenever he attempted to converse with her parents.

Millie got up to clear the table, suggesting coffee would be good with dessert. Malcolm brought his plate to the sink.

Jolene came up behind him, touched his shoulder. "Why don't you go back and keep Dad company? Mom and I are going to have a little girl time here in the kitchen." Malcolm took the hint and returned to the table, Marty had disappeared.

Delmar tapped the dining table surface. "Have a seat Doc, what's been going on over in your side of the mountain? I imagine you're seeing a lot of those dutchmen getting married this time of year?"

"Yeah, a few, along with the fall plowing."

"How about the rabies? I heard they found another cow down in Milford County. Sure is a terrible thing. So the disease comes on pretty quick then?"

Malcolm pouted. "Well, it can vary from animal to animal. In school we were taught that once dogs develop unusual behavior, they are able to infect other animals or people for only a short time."

"How do they know that, Doc?"

"I suppose from case reports or research studies."

Delmar leaned his elbows on the table, pointing with his finger, "But if a dog bit me today, could I be *certain* that I wouldn't get rabies?"

"If the dog was vaccinated, yes. But it would be kenneled and watched for ten days to make sure."

"So you could be *certain* if he wasn't showing signs after ten days, I'd be okay?"

"Animals don't survive long once the virus gets in their brain and starts to shed in the saliva, Delmar."

"Doc, no offense, because I think you're a pretty unassuming guy for all your book learning. But them know-it-alls in colleges and government don't know it all. I learned that much from farming all these years. God damn government telling us first to plant crops, then next year pay us to take acreage out of production. The only ones who make money from farming are the bankers who keep us barely surviving between loans and the pirates in the commodity markets. They couldn't start a tractor even if there was an arrow pointing to the ignition." Delmar's red nose had a luster from the combined effects of wine and heat from the wood stove. "Farms going out of business 'cause of low milk prices, equipment breaking down that's too expensive to replace, blasted weather, and the damn banks."

Delmar revisited his glass of wine and murmured, "Just shooting craps..." Malcolm held his thoughts while Delmar slid into a nap.

Soon after, Jolene wielded an apple pie into the room. She smiled superciliously. "I see you and dad got everything sorted out."

Millie followed with a percolator of coffee. Delmar started awake and muttered, "Could use the coffee."

Warm cinnamon-laced apples streamed through cold vanilla ice cream in Malcolm's mouth, a slice of heaven.

On the way home, Jolene sniggered as Malcolm recounted Delmar's diatribe. "Dad doesn't much care for Wall Street bankers. He gets worked up, usually takes a nap for an hour after Sunday dinner and wine and then goes out to milk."

"Does he really believe some of that satellite stuff?"

Her attention loitered among the scenery out the passenger window. "We have a little eccentricity in our family."

"The men or women?"

She faced him. "Dad's brother Chub perches on one of the

more twisted limbs on his side of the family tree."

"More out there than Delmar?"

"Chub has a small place farther up the road from town than my folks' farm. Lives in a small cottage by the river's edge. He's an avid conspiracy theorist."

"UFOs?"

"Yup."

"Energy particles?"

"Covers his feet and head with aluminum foil for protection against electromagnetic energy in the ground."

"Wears that all the time?"

"Not in his house, he's got copper rods and wires staked in the ground all around his yard to deflect the stray voltage."

"And I thought my family was disconnected with reality."

Malcolm pulled Lucille's headlight switch, "Your dad has been around." Jolene slid over on the bench seat and laid her head on his shoulder. She flinched when asked, "Has anyone else ever been your dinner guest for the Delmar experience?"

"Other than high school boyfriends? One."

"Oh?"

"His name was Kye. I met him up in Vermont. He was a little bit older than me."

"Were you serious about him?"

"Engaged for over a year." Malcolm didn't respond. She continued. "I was a skier while in college and met him on a ski club trip. He was a ski pro at the resort. I went up there to stay for a winter break, it turned into a couple years."

"What happened?"

She gazed across Lucille's hood at the road signs that reflected brightly, only to dim again as the headlights swept by. She replied in a soft voice, "He was killed in a car wreck in a snow storm. I didn't have any reason to stay up there."

"I'm sorry, that must have been hard to deal with."

Jolene froze her stare out the windshield. "Yeah, it was tough."

Malcolm wandered his hand onto her knees and traced the seam of her blue jeans into her lap. She grasped his hand and dwelled on her thoughts of the past for the remainder of the trip home.

When they arrived at her place, she asked, "Come in for a little while? There's something I want to show you."

He cut the ignition. "Sure, what is it?"

She opened the passenger door. "Come on in and find out."

They entered her apartment and she directed him across the small but tidy living room to an oak table that was covered in photos, graphic art supplies and drawing paper. She flicked the light switch and he adjusted his view to a spread of colored pencil and chalk sketches of scenes from Gillstown, except for one. It was a sketch in charcoal of Malcolm, naked from the waist up; the bottom half of the paper was as yet unmarked.

Feeling self-conscious, he said "This is a surprise, both the quality of the work and your subject matter. I don't remember any modeling sessions."

She looked at him from across the table. "Drawing helps me meditate and let go of the clutter. I have a good memory for detail." She tapped the paper where the charcoal ended. "Although a refresher can be helpful."

Her hands sifted through a loose pile of photos and she sorted out one to cast in front of him. "Thought you might find this interesting."

Malcolm studied the black and white photo, which captured a scene of two men by a car in front of a lodge, actually more of a manor. The lawn was manicured and the clothes of the subjects were well tailored and stylish. There were numerous people in the background suggestive of a lawn party. There was something familiar about the scene, but Malcolm couldn't place the people or place, the clothes and Fedora hats would match the late 1920's or early '30's. "Should I know these guys?"

"Not personally, but one of their descendants, yes."

It was the car that unlocked the puzzle. "My God, that looks like the same Ford Coupe I saw at MacDougall's…it's their lodge! God Almighty, has that place gone to hell!"

Jolene tossed a copy of a newspaper article in front of him. "That picture was taken at an occasion not too long before those two '*gentlemen*' died."

"Died?"

"The younger one is Trent, or 'Trotter' MacDougal, that's his father, Jebediah."

"Shannon McMorrow mentioned this fella to me a while back."

She studied his face. "Shannon, as in the ringmaster of the Dead Bear Bar?"

"Yeah, he was filling me in on some of the local lore regarding the MacDougalls."

"Well, it seems Trotter felt a little frisky one day, robbed a bank in Salty Dog, killed a local cop and eventually brought the party back home. The MacDougall's and their hired man were gunned down along with a state trooper. The article says that Trotter died next to his car, so full of holes it wasn't acting as cover anymore."

As he scanned the article, Malcolm considered the coupe and his ill-advised foray with the dog. "That's creepy."

Jolene set out another clipping from the *News-Gazette*. "This is from 1978, five years ago. Jack Calhoun was chilling up in the state pen until not too long ago, assault and battery on his ex-wife. Must have been pretty bad to get convicted, women are generally viewed as a man's property in Galloway County. Looks like he's raring to have more fun; now that he's back home."

Malcolm nodded. "Muddy waters run deep." He looked at the picture of the lodge once again and felt as if something was slightly distorted from the perspective of the camera.

Jolene said gently, "The die was cast on this clan feud a long time ago Malcolm. Maybe stretching back to when these families first arrived in these hills."

He glanced at her and beheld the likeness of his torso. "When does that get finished?"

She walked about the table and reached down to unbutton his jeans. "We can work on that a little at a time."

CHAPTER XXX

Grey smoke slithered out of chimneys and coiled over farmhouses like sleeping serpents in the still air. A chorus of pops and claps from gunfire rang out across the valley; it was opening day of deer season.

Jonas P. Yoder stroked his driving horse's neck while the creature slurped water from the trough. Snow flurries whipped through the air as the young veterinary's spotted vehicle meandered up his farm lane.

"*Guten Tag, Tierarzt.*"

"*Guten Tag,* Jonas. Thought I'd drop by to see how the ram was doing."

"Come with me. He's got a little less appetite I think."

Snowflakes pelted Malcolm's face as he scoped his all too familiar patient. The ram's head sagged to the ground as if stricken with drowsiness. Malcolm leaned against the fence and whistled. The ram sprang to life and charged at Malcolm, butting the fence with a ferocity that cracked the bottom plank. Malcolm stumbled backwards. "Good Lord!" The ram relapsed back into a stupor, making Malcolm wonder if the sudden assault was imaginary.

"We need to put him down and get him to the diagnostic lab, Jonas."

The farmer stood in silence, his nose running from the cold air.

Malcolm asked, "Have you had any other sick sheep?"

"No."

"Hopefully, whatever is causing this problem is over."

"Shall I get my gun?"

"No, not now. I won't have enough time to get down to the lab before it closes today. I want to transport the entire animal and would rather the carcass be fresh for necropsy. I'll come by later tomorrow morning." A ponderous cloud billowed across the valley, heralding a burst of snow that soon swirled about them. Deflated, Malcolm shook his head. "Tomorrow morning, Jonas. And keep away from him."

The sharp breeze nipped at Jonas P.'s exposed hand as he leaned on the doorframe of the stable, watching the veterinary's car drive away. Two sheep and a lamb lost since the black demon had torn through his flock. The farmer lamented. How was this part of the Lord's plan for his life? And why did the veterinary drive a car, but God wanted him to drive a buggy, especially on these cold miserable days? The cows needed to be milked. His young dog anticipated his master's intent and trotted for the relative comfort of the barn.

*****

By the time Malcolm reached Dale Calhoun's farm, rays from the setting sun funneled from a small crack in the grey ceiling on the horizon. For a few moments, even the somber, lifeless fields acquired a warm luster. The light streaming from the Calhoun's kitchen window drew Malcolm towards the house. The floor of the porch creaked louder than usual, complaining of age and the cold. Before Malcolm could knock, the door opened and an Amish boy, bundled in a brown thigh-length coat, stepped out into the brisk air.
"Hello, young man."
The boy smiled, left the porch and made for the farm lane.
Dale appeared in the doorway. Malcolm asked, "Isn't that the kid that I saw in your house back in the summer?"
"Yeah, Sammy Yoder, Jonas P. Yoder's boy."
"Jonas P.'s?" Malcolm turned and measured the little figure's progress as daylight faded.
"Gotta hurry back from TV time before the evening chores; nightfall isn't too far away."

174

"He's got a bit of a hike."

"Fifteen or twenty minutes if he cuts through the fields."

Dale's kids were yelling and carrying on in the house.

Malcolm cringed as Sally screeched to regain control of her brood. "Where's the cow, Dale?"

The farmer pointed to the barn. "In the stable, big white cow near the east end. I'll get my coat and be out in a minute or so."

Malcolm's eyes rested on a deer carcass hanging by the barn door. "Someone got their buck."

"I took an eight-point early this morning. He'll help fill the freezer for winter."

"I see you already removed the head for the trophy mount."

"Jack likes to do that kind of work."

"Okay, I'll get my stuff and see you in the barn."

*****

Malcolm was attending to one of the larger cows in the herd as Dale tromped into the stable.

"She's a little pokey, Doc."

"How long?"

Dale narrowed his eyes and looked upwards. "I noticed her this morning after getting my buck."

Malcolm shook his thermometer, placed it in the cow's rectum. "Big girl."

"Probably over sixteen hundred pounds."

Malcolm gently stroked the udder. "How long she been fresh?"

"Oh, jeez Doc, a while. You've already called her safe with calf."

Malcolm listened to her chest and abdomen. "She's got a slight fever, any diarrhea?"

"Not that I've seen."

Passing his hand over the cow's chest, Malcolm was surprised to feel crackling and popping under the skin. A push of the finger yielded a sensation of plastic bubble wrap. "Dale, I think she's got a little pneumonia."

Sally was at Dale's side. "What do we have, Doc?" Malcolm tapped on the ribs. His fingers discovered wet hair. Rubbing gently, pink-tinged pus surfaced. Malcolm rummaged for his flashlight and clippers from his bag.

Dale wisecracked, "Chest surgery, Doc?"

"Dale, can you please plug me in to that extension cord by the radio?" Malcolm handed Sally the light and the clippers buzzed to life. He probed the small fissure with a hemostat. "Dale, were any of your cows out on pasture this week?"

"No."

Sally chimed in, "Especially during hunting season, you know some of these city folks, wouldn't know a cow from a deer even in plain daylight."

"So has your herd been outside at all?"

"Well, they go outside for exercise while I clean the barn each morning."

Malcolm contemplated his options. "I think someone might have gotten a little trigger happy. Perhaps an infection brought in by a bullet. I can feel all sorts of air under her skin here. It might have come from a tear in the lining of the chest cavity."

Dale's face flushed red. "Ignorant pricks! How can you mistake a big white cow for a goddam buck?"

Malcolm probed with his hemostat, pausing to open the finger rings for grasping. He gently tugged out a small object. "Eureka. Looks to be a .22. Thankfully it got lodged into a rib."

Dale and Sally glanced at one another.

Malcolm studied the artifact. "I didn't see any signs of anemia in her membranes. If we can get on top of the infection with antibiotics, she might pull through. She's a big girl and the impact may have been dampened by her mass, or maybe it was a ricochet. Let's go get the drugs for her in my Scout."

Dale sulked behind Malcolm. "Ignorant bastards."

"Makes me just plain sick," Sally added.

Malcolm picked through Lucille's assortment of boxes. Dale watched his efforts, his eyelids partially closed. "You know who did this? That son of a bitch up the road."

"Dale, I came across a similar case last year, someone's cow shot during deer season. Folks do get a little bit of buck fever around here."

"He's going to get his."

Sally clasped Dale's arm. "Now hon, just calm down. Doc may be right, a stray bullet during buck season. You know we can think of at least three places over the years where folks had their house or barn hit on opening day."

"Be thankful it wasn't one of you, Dale." Malcolm said.

Dale's eyes didn't soften.

*****

Malcolm made a slight detour on the way home and parked by the township building. Cal was sitting at his desk, a *Field and Stream* in his hands. "Hello, Doc, how are things?"

"Fine Chief, and you?"

"I'm well, thank-you. Did you get your buck today?"

"Don't hunt much."

Cal placed the magazine on his desk. "Oh?" A sonorous gurgle rumbled from one of the jail cells. Malcolm took gander towards the distraction.

Cal explained, "Someone needed a little nap time from relying a little too much on Jim Beam as a hunting partner."

"That's a good mix, firearms and alcohol."

Cal folded his arms. "Something on your mind, Doc?" Malcolm relayed the findings at Dale's farm in a low voice. Cal stroked his chin with his fingertips. "Folks do get a little carried away at first light on opening day, everyone's itchy to get their buck."

"True, Chief. But Dale's cows would only have been out during barn cleaning, that's after milking while he's having breakfast and in daylight. The other thing, which I didn't tell Dale and Sally, is that wound was probably a couple days old, before buck season started today."

"I appreciate the information, Doc. And I appreciate your discretion in releasing some of the details." The chief's voice carried a pedantic tone.

Civic duty complete, Malcolm made to leave and allow the chief to revisit his reading. He paused to look at a bulletin board near the door with pictures of a couple missing dogs in the neighborhood. "One would get the impression you have more canine than human fugitives around here, Chief."

"Keep your eyes open for them if you can, Doc."

Malcolm scanned the physical descriptions. "I'll do that."

*****

Malcolm's head sank into his pillow. He missed the soundscape of summer nights that paced his release into slumber. A series of unconnected thoughts kept chafing his sleep. MacDougall and his damn black dog, so much strife and misery from one dog. One dog—there was only one dog at Jonas P.'s heels today! He tumbled out of bed and poured a finger of Oban, unwilling to turn on the lights. There were too many pieces to fit together. Massaging Precious' neck and back, he wondered, "Where was Kip, girl?" Precious licked his wrist.

*****

Malcolm leaned against Lucille, arms folded. His news for Jonas P. Yoder, standing next to him with his hands in coat pockets, had not been good. The ram had tested positive for rabies. Malcolm recalled the debate at the office earlier that day that had set the tone for his present visit…

"Jonas P.'s dog is a potential health risk," Malcolm said. "What if he has rabies, too?"

Paul shook his head and looked at the ceiling. "That's even assuming the dog has rabies."

"With everything that has happened on that farm, why else would such a loyal collie as Kip run off?"

"So every time we hear of a stray dog, we'll jump in and diagnose rabies?"

Malcolm grimaced, thankful that Paul was not his employer.

Lois mediated in a quiet voice, "Couldn't we alert the folks in that area to keep an eye out for him?"

Henry added, "I agree that this has been a strange turn of events, but we can only do so much given the circumstances."

Malcolm asked, "What of the public health department?"

"You can let Jonas P. know of his possible exposure, that's all you can do. It's up to him to decide what to do about it."

"On account that he's Amish, especially a white topper; that probably means he'll do nothing…"

Still agitated, Malcolm scuffed a boot in the hard clay pack of Jonas P.'s barnyard. Swallowing with difficulty, he murmured, "I know this is hard Jonas, but if you see Kip, and he doesn't seem right, I think you should put him down, let me submit him to the lab. Since that ram was positive for rabies, it's possible Kip is infected as well." Jonas P. straightened his back, his brown hat practically oozing moisture in the damp air.

Rachel appeared, holding a pint jar of strawberry jam. "Thank you for the books."

Malcolm beheld the little piece of summer in his hands. "Thank you."

The girl melted away. Jonas P. said, "She was close to Kip."

Malcolm contemplated the farmer's daughter as she receded into the house. "Jonas, did she or any of your other children spend much time near Kip before he left?"

"I don't think more than most days."

Malcolm tried hard not to flinch. This isn't going well. "Have you considered contacting your doctor or the health department?"

"Don't much need help, veterinary. The Lord provides."

Malcolm thought. He's going to provide a bucket load of more problems the way this is starting to line up. He scanned across the season-weary pasture with thinning grass. "Something tells me Kip isn't coming back."

Jonas P.'s eyes cast upon the ground. "*Ja*, maybe."

"I still would suggest you contact the county health department for the sake of your family."

Jonas P.'s blue eyes welled up. A wave of discomfort encircled Malcolm. "I'm sorry, Jonas." Malcolm withdrew to the Scout in a pervasive drip from the low clouds hovering over the valley, which became a soggy drizzle, which opened into a steady miserable rain on a rising wind. For the duration of the day, the foul weather and his temperament were well matched.

WINTER

CHAPTER XXXI

Snow pellets ricocheted off the windshield as Malcolm rubbed his hands with a dab of udder cream, soothing his cracked skin. Precious sniffed at the lemon scented balm. Malcolm deposited a small dollop on her nose. She wrinkled her muzzle and wiped a paw across her face.

His free hand cast about under the seat and retrieved his other winter indulgence, an Aladdin stainless steel thermos. Steadying the steering wheel, he held the cup with one hand while pouring hot chocolate with the other. Filling the cup half full, he propped the thermos between his legs and searched for the screw cap, which had tumbled between his legs. Precious yipped two beats and Malcolm swerved back into the right lane, lifting his hand from the wheel in a quick salute to the passing feed truck driver. The first sip of hot chocolate nearly scalded his throat. Schnapps would be an improvement. For that matter, so would the couch in front of his fireplace, preferably with Jolene.

Still balancing the cup and wheel in one hand, Malcolm called Lois on the radio. "Nothing new here, Dr. Malcolm, at least up your way. Oh, wait, a Dr., uh…, Teesdale called. She says she's an epidemiologist at the state veterinary college. Has some sort of interest with our rabies problem in the valley and wants to talk to you about some of the cases."

Malcolm held his reply until he negotiated passing a black top buggy. The driver and his wife were huddled under a blanket, as if carved of stone. That looks cozy.

"Malcolm, you there?"

"Yeah, I'm here. Did you get a phone number?"

"Yupper, maybe you can try to get her tomorrow."

"We'll see how the day goes, thanks, Lois."

"Stay warm." After a hiatus, Lois' voice returned. "I saw that stromboli wrapper in your car the other day. Awful lot of grease in those things."

"I'll eat something healthier today, perhaps fried chicken."

"Oh my God, I'm working with children. How's my favorite girl?"

Precious responded with a protracted, "Ruff!"

Malcolm released his thumb off the mike button. He regarded his travel companion. "Next time I'll hand this to you, then I won't have to listen to the dietary lectures. You were pretty eager to help me eat the pepperoni from that stromboli."

He wondered about the call from the university, a research survey perhaps? Was it related to other rabies outbreaks? Savoring his hot chocolate, he cranked up the stereo so John Lee Hooker could let it all out, leaving a trail of ivory powder roiling behind the Scout. The day list was long and the daylight short, a return home before sundown was doubtful.

*****

Even in the lackluster light, Jonas P. could see that something was amiss. The shed door was open, and the stiff wind banged the door into the planks on the outside wall like a boxer training on a punching bag. He lifted his lantern and cased the ewes and yearling lambs, huddled in a dim corner furthest away from the entrance. Jonas P.'s accusing glare sought the alleged perpetrator of this most heinous felony of animal husbandry, leaving a gate unlatched. His son blanched and professed his innocence. Farmer and son emptied pails of grain into the wooden feeder. Oddly, rather than jockey for position to eat, the sheep kept their distance.

Snatching the lantern, Jonas P. left his son in the shed, bracing against the biting air stinging his cheeks. The snow at his feet was trampled into a muddy slush, with a distinct furrow leading behind the shed. Raising his lamp, his bare hands numbing from the cold, Jonas P. followed the track to find a prostrate ewe. Her vacant eyes were blurred into a glassy film and the fleece about the gash on her throat was stained a dirty red-brown. Anger welled up within Jonas P. Yoder. The *schwarze Hund* had somehow returned. No longer would the rifle remain in the barn. Now it would lean against the doorframe inside the house, loaded and ready to fire at trespassers of any kind.

CHAPTER XXXII

John MacDougall scratched the stubble on his face and helped himself to another dose of oblivion. His love affair with whiskey had been a lifelong commitment fostered in a family with a devout affliction. Intolerance and anger were his drinking partners. He had his fill of the Calhoun brothers and their buddies in the Dead Bear Bar, and that self-righteous young vet, the nosy asshole who didn't know shit about life here in Galloway County. But most of all, he despised the dutchmen. Always stuck to themselves, wouldn't speak English after all those generations in America. Wouldn't go fight for their country, not like his kid brother Archie. Archie came back from Vietnam a wreck of a man while the dutchies stayed at home and grabbed more land. They had always wanted the MacDougall land, piece by piece that had taken it away from his family over the years. Always more than willing to lowball an offer for a parcel when MacDougall fortunes were in dire straits. Acting like they were fucking holier-than-thou, sitting in their goddam buggies, all the while gossiping and carrying on with their superstitions and witchcraft.

Archie went to Nam as a young man full of grit; came back decorated with Agent Orange, jungle nightmares, a messed up back after a bad jump from a Huey, and a mind full of smack. Archie was never right from then on. When Archie was in one of his moods, he would whisper to his elder brother that he was 'going to hunt 'Charlie' that night. Cloaked in loose-fitting combat fatigues, face painted, he would drift into the woods beyond the clearing at sunset, not returning until dawn to awaken his older brother with a rap on the kitchen window.

MacDougall was still haunted by Archie's feral eyes leering through the panes, his shoulder length brown hair tangled and matted in mud. Archie would clasp his brother's hand as they met on the back porch and boast, 'I got me one'. A raccoon, opossum, or even the occasional dog or cat, would be hanging on a fence rail, gutted like a deer, by the family plot, a sacrifice for the ancestors. Despite MacDougall's careful admonishment and disposal of the carcasses, invariably another quarry would appear after the next 'patrol'.

John MacDougall had sought help for his broken brother, but the overworked drones at the VA hospitals were useless. After all the patriotic drum beating, suitcases full of cash hauled away by defense contractors, and over 50,000 dead, the Pentagon, politicians, and most of all, the public wanted to conveniently forget. Nobody wanted to piss more money down the rat hole of shattered souls where his brother dwelled. He was one of the abandoned casualties of the empire, day after stinking day, year after stinking year, MIAs in their own land. And all the while, the fucking dutchmen sat in the valley, unwilling to serve like his brother. Goddam dutchmen, worse than the Viet Cong.

Archie never ate inside the lodge, preferring dried goods such as jerky, or apples in his 'command post' in the barn. His favorite was peanut butter and bacon sandwiches on white bread. Sometimes, under the full moon, Archie would sit by a small fire and chant a language that MacDougall didn't recognize, if it was a language at all. Lester, MacDougall's son, afraid of his uncle, left home for Gillstown, where he got a job as a mechanic. Unlike his son, MacDougall couldn't cast Archie aside. Not his own kid brother, not from the blood and the land of their forefathers.

Local rumor had it that there were missing pets and sightings of stalkers in the night. But John MacDougall wasn't intimidated, not by the Calhoun boys, the police, or anyone else. No MacDougall had ever backed down from doing right by their kin.

In the end, a confrontation was unnecessary. Archie packed up a rucksack and announced he was going "to hang with my army buddies." MacDougall figured his brother needed to crash where his habit could be more easily fed, down in Philly or some other shithole city. The homecomings became less frequent and the duration of the absences grew like a spreading poppy flower in the warm sun. It was just a matter of time until the fragments of his brother's mind would ride one last high and not return back into this world, free at last from the demons in his head.

MacDougall pushed himself from the table, the bottle of whiskey drained. Staggering across the floor, he yanked a replacement from a cupboard and slumped back to his chair. Sleet and snow had collided against the sashes of the lodge for the better part of the night. Now, the bleak chill of the oncoming dawn reminded him of his lack of sleep. Worn out from the night's binge, he clamped his eyelids. Hardly perceptible above the wind, a tap rattled the back kitchen window. An empty tumbler slipped from his hand and rolled across the table as he sat upright in his chair. The glass pane strummed again. John MacDougall managed to get to his feet and charged the door, knocking a chair over in his haste.

CHAPTER XXXIII

Jolene savored her coffee, fixated on Malcolm's choice of clothing as he rooted about in the refrigerator. A purple bathrobe with his initials over the pocket and plaid pajama bottoms, apparently a gift from his mother. By God, she must be colorblind. She turned her attention out the back window and took notice of a moving shape beyond the dead flowers in the garden. "What's that dog doing out in your yard?"

Precious popped her paws onto the windowsill and barked in a high pitch.

"Malcolm, what is going…" Jolene grinned. "You have a visitor."

Malcolm swung open the back door. "Go on! Get out of there!" A hefty chocolate brown lab lifted a hind leg by a hedge, satisfied his need and wiggled his mass through the back corner of the fence. Precious wagged her tail, whining by the door.

Perplexed with his oversight, Malcolm grumbled, "Some critter must have dug a hole in that corner when I replaced the fence boards."

"Have you seen him before?"

"No."

Jolene considered Precious. "I think she has. He was an intact male, by the way."

Malcolm eyes widened. "It couldn't be."

"I thought she looked a little pudgy this last week or so."

Malcolm bent over Precious and palpated her underside. She licked his face. "It couldn't be." He stood up and rubbed his forehead in the palm of his hand.

"Is she spayed?"

"The folks at the animal shelter said she was."

"So unknown family history?"

He folded his arms. "Not unknown anymore."

Jolene laughed and rubbed Precious' back. "You're going to be a mom, girl!"

"Swell, the local vet's dog pregnant because she isn't spayed. The very thing I always preach about."

"Veterinarians who live in glass houses…"

He stared through his reflection in the window and out across the back yard. "First, I'm going to fix that fence."

"I'm going to shower", she responded airily, leaving Malcolm by the window and Precious ensconced by the door.

"Oh, by the way," he called out, "how about dinner on Christmas Eve with my folks? They're threatening to visit for a couple of days."

She spun about. "Your parents are coming? You didn't tell me."

He examined his mug. "I need another splash of coffee."

*****

Malcolm took a big chomp off the top of the Christmas tree. "They say it's going to be a terrible cold blast, right on Christmas day."

"Much more of a pain in the ass than that cold snap that came through last week," Dale complained as he fingered another green-iced cookie from a pile on the kitchen table. "Tough on the calves, trying to keep the water from freezing, tractors won't start, drafty barn…"

"Drafty house," Sally sighed.

Malcolm washed the cookie down his throat with coffee. "How cold is it supposed to get?"

Dale exhaled, "Colder than a well digger's ass, Doc. Below zero anyways, That's when the fires will start. Everyone gets the wood stoves going full tilt. All those chimneys that need cleaned."

Malcolm grasped his cup tighter to feel the heat, ruminating on his emergency duty for the holiday. "Hate to be one of you guys in the fire department in this weather."

"Hard on everything and everybody," Sally nodded as she warmed up the coffees all around.

"What about that dutchman's sheep with rabies, Doc?"

Malcolm looked at Dale over his mug. "Not much since then. I haven't heard much about MacDougall's dog as of late, either. Have you?"

Dale looked towards the ceiling and shrugged. "Nope."

"Nothing?"

"Nope."

Malcolm wondered if the MacDougall-Calhoun feud would ever end. Not likely. "Well, thanks for the break, guys. I better get going."

Sally offered the plate. "How about another cookie for the road?"

Selecting Santa trimmed in red and white, Malcolm said, "Much obliged." He halted by the door. "Say, you wouldn't know who might own a large chocolate lab, would you?"

"One of those lost dogs people been wondering about, Doc?"

"No, not missing at all."

Dale creased his forehead. "Don't know of any."

"Thanks." Malcolm stepped out onto the porch and surveyed the hibernating ridges on the horizon. *Sub-zero on Christmas? I hope it'll be a quiet day.* Time was growing short for food shopping for his parent's arrival. He smirked. *Farm girl meets Chicago urbanites. Better get some wine, a little lubrication will be in order.*

*****

Blasting mayhem through the northern plains, the blizzard surged into the Great Lakes, a juggernaut of wailing gales and bitter cold. Malcolm was chopping celery when the phone rang, providing a welcome distraction from the dire weather forecast on the radio.

"Malkie? How are you dear?" Only one person on the planet called him by his kid name.

"Hello, Mom. I was starting to get things ready for tomorrow."

There was a brief pause. "That sounds divine. We *so* much want to come and were ready to go, but the weather out here is unbelievable."

"We're in for the same weather tomorrow night, perhaps as much as ten inches. And I guess the winds will be pretty wild the day after."

"That would be on Christmas day?"

"Yes, the roads will be a mess."

"That's what we heard. Do you think it might be difficult to travel then?"

"No doubt."

"And we wanted to get to your sister's place later in the week, you know, to see your niece."

Malcolm sensed he should offer the out. "Well, traveling will be next to impossible for the next several days, let alone dangerous. And I'm on call anyway. Why don't you wait until another time, when we have better weather?"

"You're a dear. Are you sure you wouldn't mind? We really want to see you."

"It just makes good sense. I'll be fine."

"We hate leaving you by yourself on Christmas."

Malcolm considered the benefits of no last minute house cleaning, or political diatribes from his father, as a positive development. "It's my first Christmas on call. No big deal. Jolene will probably still come over tomorrow."

"We'll have to meet her sometime. Could she come and visit us?" He thought, Dear God, not even the neutral territory of his own house for the first meeting.

"We'll find an opportunity." Malcolm knew better than to commit without consultation.

"We'd absolutely *love* to meet her."

Malcolm listened politely to a discussion on family affairs. The conversation ended with mutual wishes for a Merry Christmas. The chopped celery lay before him; spurred by the image of a quiet dinner together, the preparations for the meal continued with vigor. Parents or no parents, wine would still be in order.

CHAPTER XXXIV

Menacing clouds thickened over the ridges, slowly dissolving the low-lying sun into a grey-yellow film. Nursing his aging Ford along the Shaneytown Road, Chief Cal's mood matched the weather. His right knee forecasted that something harsh was on the way. He pondered what the current temperature was at his property in Florida. Thirty-three years of refereeing the law in Salty Dog and the surrounding environs of Kilderry Township had worn his patience thin, especially for situations like this afternoon. Cal came up over a rise in the pavement to find a volunteer firemen's pickup parked behind Tommy's car. Cal scowled. *This will get around the rumor mill, the last thing I need.* He pulled behind the truck and cut off the engine.

Tommy popped out of his cruiser and shielded his face from the wind as the chief lifted himself through the car door. Cal looked at Tommy with concern. "Shouldn't we be thinking of traffic control for the rubberneckers?"

"This drizzle is cold," Tommy explained. Cal lowered his head, took a breath and reminded himself that Tommy was his sister's boy. "Why don't you show me what we've got?"

Tommy turned, his yellow rain gear billowing with each gust. "Well Uncle Cal, Harlan and Butch here were driving back from hunting, when they came across this poor little guy on the road."

"What poor little guy?"

Tommy led Cal to the shoulder of the road, joining a couple of young men in orange vests. At their feet lay a mixed breed terrier. Its white fur was stained red about the neck, its lifeless eyes staring into the barren field beyond. The rain spat at Cal's face as he squatted down and lifted a front arm and let it drop. He clenched his jaw. Damn, this looks to be one of the missing pets.

Tommy leaned over his shoulder. "The poor thing has got a collar and tags. These fellas thought we might be able to find out who he belonged to." Cal straightened back onto his feet. Tommy continued, "Even if he was hit by a car, this is the second one in the last couple of weeks around here, and with that tear…"

"TOMMY!" Cal roared. Lowering his voice, the chief advised, "How about you find a blanket or towel to wrap around the body?"

"But Uncle Cal, don't you think…"

"Now!" The two hunters regarded one another. Cal shifted his stony gaze on them. "You fellas call this in?" They nodded eagerly. He squinted against the weather, scanning the perimeter. "Well, we'll take it from here. Thanks for letting us know."

Tommy returned and gently scooped up the dog in his arms. "I'll carry him to your car, Chief."

Cal limped after his nephew to the relative comfort of his cruiser, thinking, even on a rotten day like today, this news will get around town like a wildfire in a drought. Cal opened the trunk of the car, pointed for Tommy to lay the lost soul gently onto the floor. A wicked wind reached across the open field and lashed sleet into their faces.

*****

Malcolm bent through the back hatch of Lucille, his lower back muscles complaining from their participation in paring out a foot abscess on a lame cow. Strolling out of the barn, John Y. Peachey quipped, "You know *Tierarzt*, maybe I should get my cousin, Josiah Y. Peachey, to do my pregnancy checks."

"How so?"

"He's another of Zephaniah Y.'s sons. Hangs a wrench on a chain over the cow's back." John Y. circled his hand, palm down. "It moves when a cow is pregnant."

194

"That's a good trick. Where did he learn that magic?"

The grin vanished from John Y's face. "My uncle has some learning in the hexings, don't you know."

Malcolm scratched his head. "You don't believe that do you?"

John Y. remained uncharacteristically silent. Malcolm said, "Are you telling me he sees himself as some sort of sorcerer?"

John Y. turned away and faced the east. "Wind picking up, clouds swelling over the mountain, I think old man winter visits us soon."

Malcolm tried to delay thinking about reality. "What's for Christmas dinner?"

John Y. patted his girth. "We eat goose for Christmas and pork on New Year's Day."

"Your goose?"

"My brother Eli raises them."

White specks flitted into Malcolm's hair. John Y. looked about. "It is here."

"Merry Christmas, John."

"*Fröhliche Weihnachten, Tierarzt.*"

Lois was off for the afternoon; Henry answered the radio. "No new calls, Malcolm. Looks like the bad weather is coming this way."

Malcolm watched gritty snow dance across Lucille's windshield. "Yeah, so it seems."

"Hope all is quiet. Take your time on the roads."

"How about you, Henry? Travelling to your daughter's place?"

"Leaving soon, only an hour away."

"Enjoy your holiday."

"Thanks, Malcolm, and Merry Christmas to you as well."

He cut across the valley so as to stock up on supplies at the vet clinic. Passing Nahum Y. Peachey's farm, the billboard of religious imperatives stated, ***Flee fornication. I Corinthians 6:18***.

Malcolm wondered if that included his surly bull?

*****

Ted Litwiller, assistant store manager of the Save-More grocery store, wanted to finish the day and hurry home. Christmas Eve was upon him and he figured on closing the store early. He had yet to caravan his family over to his in-laws and he disliked driving in snow. It had been a hectic day; there had been a rush on food staples on account of the weather forecast, in addition to last minute shopping for Christmas dinner fixings. Two of the remaining employees were already cleaning up the aisles and the back rooms. The store was otherwise deserted except for a cashier at the only open register and a lone customer; a disturbing and peculiar customer.

The patron was wearing a tattered dark sweatshirt covered by an orange hunting vest. Below he wore dirty white painter's pants and two leather boots, neither of which matched in color or style. His hair and scruffy beard lay about his face, lending a flair of vulgarity. But what got Ted's attention even more were the items that the figure had collected in his shopping cart—bulging over the brim with loaves of Wonder bread, Jiff peanut butter, and bacon. From his office, Ted watched as the man pushed the cart towards the checkout lane, teetering every few steps so that he had to latch onto the cart handle to steady himself. The man was either drunk or off his rocker.

The vagabond arrived at the checkout and started to fling his booty onto the automatic belt. The clerk was a high school girl and was clearly intimidated by the presentation. A loaf of bread sailed off the belt and bounced onto the floor. The man cried out, "Son of a fucking bitch!" He stomped on the hapless loaf and kicked it, sending it sliding into the magazine and gum rack.

The girl drew back, unable to move. The man looked at her and snarled, "What the hell are you looking at? It's just fucking bread; ain't like I'm kicking nobody's dog. Get on with it!"

Mr. Litwiller came to her rescue. "Is something the matter here?" He stopped and took a closer look at the ragged shopper. "Mr. MacDougall?"

"Hell yes, something is the matter, I've got to haul this stuff back home before the storm comes and this little shit can't figure out how to do her goddam job." MacDougall's breath reeked of alcohol.

Alarmed, the manager said, "There's no reason to get upset about this, we'll…"

"Who the hell is upset? I only want you bastards to ring up my grub!"

Mr. Litwiller turned to the girl. "Please ring him up, Nora." He circled around and took up a position to bag the food. There was enough to fill nearly six brown paper bags.

"Careful with that bread, don't want it squished," MacDougall said.

As the total was rung up on the register, MacDougall blurted out, "I'll take two of them Slim Jim's and two cartons of Marlboro's. Nora looked at her supervisor, wondering if she needed to cancel out the sale and add the new items. He shook his head vigorously.

MacDougall reached into a pocket and pulled out a handful of ten and twenty dollar bills and tossed them in front of the girl. He then jammed his grocery bags one on top of another in the cart and wheeled towards the door.

Nora called out meekly, "Sir, do you want your change?"

The departing shopper made a motion as if whipping thin air and shouted, "On Dasher and Prancer and to all a merry good night." The hum of the automatic door was the only sound as he and the cart vanished into the parking lot. Mr. Litwiller hustled to lock the door.

*****

At home, Precious bounced across the backyard in the deepening snow; her gait having become a swagger from her extra burden. Watching her, Malcolm sighed. Should never have assumed the shelter knew her history. Bonnie Prince Charlie sprawled across his favorite armchair, upside down, one eye open. With the approaching weather, his highness was likely to stay for an extended

197

visit. Miss Violet had left a plate of Christmas cookies and a card on the kitchen counter.

The snowfall expanded, driven by a wind that searched for any small chink in window sashes or door frames. After washing off the day's labors under the shower, Malcolm pulled on jeans and a sweatshirt and deposited a Tijuana Brass Christmas album on his ancient turntable. Trumpets blasting, he prepared dinner, sipping cabernet.

A couple of album sides later, with darkness approaching, he went to turn on white lights draped on the crab apple trees in the front lawn. The headlights from Jolene's truck highlighted his fumbling with the connection between extension cord and outside socket. Shuffling through the snow to greet her, Malcolm pointed to the garage door. "Why don't you pull your truck inside the garage?"

"What about the firewood that's lying all over the floor?"

"I stacked it along the wall the other day."

"I'm impressed." Jolene eased her truck next to Lucille. Malcolm secured the door, snow seeped inside his moccasins and burned his skin as it melted on his feet.

"What's in the bag?"

"Homemade cranberry bread for tomorrow morning and something for your stocking."

"Coal I should think."

"You deserve it. What's for dinner?"

He hugged her. "A nice red wine."

"Good start."

"And on the food side, wild rice stuffed into Cornish hens."

"Cleaning, cooking, you're becoming an industrious housekeeper." Malcolm lifted the bag from Jolene and returned to the kitchen. Precious greeted them by poking her nose at the bag.

Lifting the lid off a pot, Malcolm inspected the progress of the rice. "If all goes well, we eat in about half an hour."

"Can I start the fire?"

"By all means."

"Where in the world did you get this schmaltzy music?"

"Music stores that specialize in schmaltzy music."

Before long, a red-orange sprout flickered in the fireplace and Jolene cast herself onto the couch. "Too bad your folks didn't come up."

He joined her with a glass of wine in either hand. "Probably just as well; with me being on call I have to limit my partaking of Christmas spirits." Bonnie Prince Charlie stalked the spreading warmth in the hearth, reclining on the carpet three or four feet away. "I don't miss the family much over the holidays anyway."

"How can you not like Christmas?"

"More stress than relaxation. My grandmother and great aunt hold holiday family court, passing judgment about our life choices, helps bring forth the Christmas cheer."

"Charming."

"Uncle Fred gets pretty hammered, which is a regular occasion, and contributes to the annual celebration by proclaiming that our country is being overrun by teacher unions, Jews, coloreds, and Mexicans." Malcolm imbibed. "And my sister and her seedy husband can always be counted on to spark tedious banter over how my divorce was avoidable, if only I would have remained in Illinois and taken a more respectable job, etc., etc."

"Well, why didn't you?"

"What, and miss the winter splendor of Galloway County?" His hand coursed through her hair. "So what's up with Delmar and Millie?"

Jolene shook her head. "This storm will make a mess of things on the farm. I'm going to have dinner with them tomorrow afternoon. You're welcome to come along."

"Let's see how the day goes, I can't be that far away from the phone for so long." She looked at him with a sideways glance. Malcolm shrugged. "It's my turn at bat."

She rubbed his knee. "I know."

Reluctantly, Malcolm struggled up from the couch. "I have to check the birdies in the oven."

Over the meal they chatted over memories of daredevil rides on sleds, skating fiascos, and outrageous snowball fights. Afterwards, the warmth of the fire drew them back onto the couch.

Stretching her legs fully along the cushions, she said, "You work hard at trying to be a non-conformist."

Malcolm smiled and her eyes softened as he drifted a finger along the curve of her waist, circled around her navel and descended on her midline. Arching her spine, she lifted her sweater to reveal her torso, shimmering in shades of amber from the glow of the fireplace. Jolene clutched at his shirt.

A knock rapped on the front door. Malcolm cast a baleful look at the intrusion as Precious waddled over to the foyer and barked nonchalantly.

"What in the hell?"

The rat-a-tat-tat echoed again.

Jolene arms went limp. "Aren't you going to get that, *Doctor*?"

Malcolm grunted, pushed himself from the sofa, and brushed his hair. "What fool would be out on an evening like this?"

CHAPTER XXXV

Malcolm opened the door in bare feet. Chief Cal held a bundle in his arms. His eyes lingered on his host just enough to raise Malcolm's awareness of his cavalier appearance.

"Hate to disturb you being Christmas Eve and all, Doc. Wanted to get here earlier, but got called to a couple of fender benders, it's going to be that kind of a night."

Malcolm beckoned his visitor through the portal. "Sure Chief, come on in."

"Got something a little out of the ordinary I'd like you to look at."

Malcolm led the way into the kitchen and popped the light switch. Miraculously, Jolene was standing there, fully dressed. She caught Malcolm's eyes with a smug look.

Cal entered. "Good evening, Miss."

Jolene asked, "Care for a cup of coffee?"

"Thank-you, no. This got called to my attention earlier today, on the edge of town."

Malcolm waved his hand towards the table. "Lay it out here, Chief." Cal gently unfolded the blanket. Precious sniffed the bundle and lowered her head. The body of the little terrier came into view.

Malcolm recoiled subtly as Jolene exclaimed, "Oh my God!"

Chief Cal said, "I'm sorry to say that I may know the owner."

Malcolm recognized the dog from one of the flyers about town. He gently stroked the creature's back and neck. "Poor little guy."

"I wanted your thoughts on the most likely cause."

Malcolm regarded Cal, then Jolene. Gently, his fingers rolled over each leg and massaged the abdomen and rib cage. Turning the body over, he examined the face and mouth and rubbed the skull. Finally, he probed the throat. The trachea was exposed and severely lacerated.

"I imagine you figured this one out, Chief. Massive trauma to the throat. I would think some sort of wild animal."

"Is it possible he was hit by a car or dragged, so his throat was torn open?"

"Doesn't seem likely. His limbs and ribs are all sound, no fractures, abrasions…"

"No other unusual wounds?"

"Without cutting him open for a necropsy, I can't be sure of internal trauma."

Cal flinched. "Don't want to bring him back to his owner like that. This is bad enough. How about the size of the attacking creature?

Malcolm shook his head. "I don't think it was an animal attack, this is a clean cut, like with a knife. And besides, there's no soft tissue damage that would indicate a bite wound."

Rubbing his chin, Cal's voice drifted. "You don't say?"

The three meditated in a circle about the table, an impromptu wake for the victim. Malcolm suggested, "If you're going to take him back to the owner, let me sew up his skin, make him more presentable."

"That would be appreciated."

Precious lay with her head between her legs, her eyes fixed on Malcolm. Her plump abdomen expanded with each breath. As Malcolm finished the patchwork suture, Cal folded the blanket over the body and nodded to Jolene as he made his way to the door.

Jolene asked, "Where did you find this poor creature?"

"Up on the Shaneytown Road." He halted as Malcolm pushed on the door. "I would like you to keep this to yourselves. Have a Merry Christmas."

Disregarding the glacial air coming in through the door, Malcolm replied, "You, too Chief." Closing the door, Malcolm paused for a moment. "That's a nice Christmas Eve present for the dog's owner."

"What do you think is going on?"

"Damned if I know, something depraved."

"No idea at all?" Driving a finger into his chest, she mocked him. "That's rare for your Mr. Science mind not to come up with some sort of a guess." She looked over her shoulder towards the fireplace. "We have some unfinished business."

CHAPTER XXXVI

Malcolm bobbled the phone in an autonomic hand. "Hello?"

A youth's voice asked nervously. "Is this the veterinary?"

Malcolm responded in with a rough gargle. "Yeah, this is Dr. Cromarty."

"John Y. Peachey has a cow down trying to calve."

Avoiding the impulse to answer, "Well good for John Y.", Malcolm said, "John Y. Peachey on Rocky Creek Road?"

"Yes, John Y. Peachey."

"Okay, I'm on my way." Malcolm's hand slid away from his ear and dropped the phone onto the floor. Hovering on the brink of consciousness, Jolene's warm body seduced him to stay under the sheets and return to a cozy nap. Precious whimpered and doused his nose with her tongue. An eye blinked at the alarm clock—four-thirty. Damn, that's rude. With heroic effort, Malcolm pulled away from Jolene's gravity and slouched on the edge of the bed. A fierce howl rattled the storm windows. Malcolm shivered at the thought of driving Lucille. Wonder what the roads are like?

A hand palmed the small of his back. "What do you have?"

"Sounds like a calving with a milk fever. And a very Merry Christmas."

"Who is it?"

"John Y. Peachey, up on the Rocky Creek Road."

After a momentary silence. "I want to go, too."

"That wind sounds more obnoxious than a frat boy on cheap booze. Hang tight here in the nest."

"I've seen my share of early morning calvings, you know."

Malcolm sorted out a plan in his stupor. "Well, if that's the case,

I gotta find you some long johns and layers of warm clothes.”

“I’ll go start some coffee, I *need* coffee.”

He turned on a light.

“Oh,” she exclaimed, “Now that’s obnoxious.”

The wooden floor chilled Malcolm’s feet as he explored about the room for long sleeve shirts and pants. Finding his way to the kitchen, he peered at the thermometer outside the window. Minus seventeen degrees. He took a second look. Minus seventeen degrees…with the added bonus of a savage wind. This was going to be a two-sock morning. Malcolm girded two tee shirts before pulling on long johns, top and bottom. A sweatshirt and pair of jeans followed, then the main event, his insulated Carhart bib coveralls. Last, he added a sleeveless down vest.

An arctic blast assaulted Precious as she plowed out the backdoor and through a snowdrift. She stayed outside only a short while, tracking lumps of melting ice indoors as she sought a rug to lie on. Malcolm commented, “You seem rather mopey this morning.” He robotically filled two mugs with coffee and poured the rest into the thermos.

Jolene arrived by his side. “Coffee ready?”

Malcolm suppressed a smile at her assortment of poor fitting clothes. “Yeah, I’m going to start Lucille and coax her to warm up.” He grabbed two pairs of gloves, hats, and a couple of worn coats.

A frigid fist gripped the skin on his cheeks and freeze-dried the inside of his nostrils as he opened the garage door. Malcolm turned the ignition and the Scout convulsed into forced service. The cylinders sputtered as he inched back out of the garage and into the frozen barrens. The meager illumination of the headlights in the white-out offered only a shred of comfort. Leaving Lucille to enliven herself and hopefully the heating core, Malcolm closed the garage door and returned inside. “You ready for this?”

Jolene picked up the mugs, handed him one, a thread of steam wafting upwards.

Malcolm slurped a slug of coffee. “Might as well give the engine a minute or two to warm up.”

"How cold is it?"

"Polar bears would be okay with this."

"You mean it's not a good day for me to wear my brass bra?" His eyes widened. "What Malcolm, did I scorch your sensitive ears?"

"Maybe if we wait here long enough, John Y. will call back and say all is well, don't bother coming," he suggested.

"Are you awake or still dreaming?"

Retrieving a couple of old blankets from the hall closet, he called out, "We'll be back," to a recumbent Precious.

The cold vinyl of the bench seat caught Malcolm's attention, despite the coveralls and long johns. If he used his imagination, the air coming out of the heating vent was at least tepid. With a toss of the blankets over her legs, he advised, "Might help."

Pallid desolation engulfed them as Lucille crunched down the driveway. Howling gusts buffeted the sides and roof, reminding the occupants the Scout was an insignificant bubble in a polar vastness. They followed the snow-swept road, a caravan track through a desert laced with vortices of white powder. In some places, the track all but disappeared where drifts seeped over the road bank, causing Lucille to lose momentum and waver. "Thank God for four wheel drive," Malcolm muttered.

Jolene huddled next to Malcolm on the seat, her breath vapor barely perceptible in the anemic glow of the dashboard. He wanted to put an arm around her, but the skimming and skidding through the drifts made him reluctant to take his hands off the wheel, although he maintained a firm grip on his coffee mug.

"Except for the lack of a horse, this is kind of like a Yuletide sleigh ride," he commented.

"Except for the lack of heat in this old car of yours, that would almost be funny."

"Usually works better than this, maybe it's the extreme cold on the radiator." Malcolm fiddled with the heater controls.

Malcolm stole a quick glance in her direction. "You look very fetching in that hat."

"You've got to have the most ugly winter clothes. Get them on sale from the Tuesday flea market in Mayville?"

"I avoid that place like the plague. Would you rather I wear dress slacks and sweaters out to the barn in weather like this? The more layers underneath in the cold, the better."

She sipped her coffee. "What do you wear underneath your coveralls on hot days?"

"On farms with attractive farmer's daughters, only my underwear."

"Does that include Delmar's?"

Malcolm's teeth flashed. "There, nothing at all."

He turned Lucille off the state highway onto what he guessed was the pavement of Rocky Creek Road. John Y. Peachey's farm lane was half-covered in mounds of snow, shifting in the wind and striking at Lucille's tires like angry snakes.

Malcolm halted by the milk house. A fury screamed overhead and strafed Lucille with powder from the barn roof. Malcolm looked at Jolene and pushed open his door. The cold gnawed at his ears as he sorted equipment though the back hatch. They escaped to the milk house, a balmy oasis compared to the wilderness outside. Ice cold water numbed his hands while filling the bucket, he wished the Amish believed in water heaters.

The windows and doors of the stable were barricaded with bales of straw to resist the bitter air from seeping into the barn. Not even the heat of the cows could keep water pipes from freezing unless all drafts were blocked. Barn cats huddled together next to calves as they lay in their pens, anything to stay warm. John Y. converged on the travelers; a heavy black denim jacket fastened around his girth, evoking visions of a hefty middle aged man wearing his thirty year-old military uniform for a parade. The farmer acknowledged Jolene with a curious appraisal, then greeted them in a cheery voice. *"Guten Morgen Herr Tierarzt*, we haven't seen you for such a long time. And you brought a helper, I see."

"Good morning, John. I wish we could have taken care of this yesterday when I was here. Where is the cow?"

"*Komm.*" The farmer boomed out towards the far end of the lamp-lit stable. "Samuel, *Komm mit Herr Tierarzt, bitte.*" John Y.'s oldest son appeared from the shadows. "The others can milk, we will visit the cow together on such a fine morning."

Malcolm pondered the source of John Y.'s energy. Christmas spirit, numerous sugar cookies, special Amish holiday elixir from a still?. Whatever the source, he could have used some. Holding a lantern, John Y. led the way through the relative warmth of the stable and into the deep freeze of the back of the barn. They entered a box stall and waded through a deep pack of fresh straw to a cow that ignored their arrival, lying on her right side. The head of the calf was thrust out her back end, but only the head.

Malcolm gently cupped the head in his hands; the tongue was swollen and discolored. Deflated from what he saw, Malcolm asked, "Can she get up?"

John Y. shook his head. "No."

"Older cow?"

"I think maybe fourth calf."

Malcolm asked Samuel to place a halter on the cow. Meanwhile, he shed his gloves to open the calcium bottle, attach the hose, and jag the needle into her vein. "She's in a bit of a fix, John. The calf's head came out with the legs behind, probably that's how the poor thing got stuck. The calf is dead and with all that swelling, we'll probably have to cut the head off. It'll be easier to push back in and get the legs turned around that way."

John Y. rubbed his hands for warmth. "I was afraid for the calf when I saw the cow this morning, but hopefully we'll make her right yet."

The calcium seemed to run slower than any bottle Malcolm had ever administered. The interior of the stone walls of the barn were spotted with patches of frost and the pervasive cold needled Malcolm's fingers despite replacing the gloves. He perused Jolene's garments; she smiled as her eyes caught his. The apathetic bottle finally expired. Malcolm started the second bottle under the skin, handed it to Jolene, requesting, "Please stick her in a couple of other places."

Reluctantly, Malcolm removed his outer garments, tugged a sleeve over each arm, squatted behind the cow and tried to shove the calf head. It didn't budge.

Probing about in his pocket for a jack knife, he sliced through the calf's neck, twisted the skull and cut the head from the body. He realized the best angle to continue his task was to recline on his back. Malcolm groaned as his body heat diffused into the frozen dirt pack under the straw. He pressed the calf's neck from the birth canal and back into the uterus. His fingers groped about for a knee joint of a front leg, bent so that the foot stretched away from him and out of reach. He carefully wrestled the tangled limb for several minutes so as to not tear the uterine wall, while grasping the foot and straightening the leg. Finally, he got the leg to reach out through the cervix and a waxy hoof poked out of the back of the cow.

Now with more space in the uterus, Malcolm manipulated the second leg out of the cow with less effort than the first. Malcolm puffed, "I think we're ready to pull the calf now."

"*Gut Tierarzt*, I'm getting cold here. It's time for me to sit by my wood stove."

Jolene shivered. "That sounds nice."

Wrapping a chain about each leg, Malcolm hooked them with a pair of handles. "John, if you and Samuel would be so kind to each take one of these." John Y. and his son positioned themselves behind the cow and hoisted a handle with both hands. Malcolm reached back in the cow to cover the exposed vertebra of the calf and guide it out through the birth canal. "Okay, pull!"

The calf slid a few inches, momentarily seized at the hips, and popped out in a quick burst. Malcolm reached into the uterus for a final inspection, relieved to find no lacerations, or a twin sibling.

His legs cramped as he removed his sleeves and pushed from all fours to gain his feet. Icy air clutched his sweat-dampened clothes, stiffening the fabric as if starched. He shoved his aching arms into his vest and coat.

Malcolm scratched the cow behind her ears. "John, if she

doesn't get up until this afternoon, I should see her again. Make sure she gets offered some water right away."

"We'll do that *Herr Tierarzt*." After a quick glance from his father, Samuel went to search for a bucket.

Back in the milk house, Malcolm took off his coat and anesthetized his arms and hands with betadine and ice water from the faucet. John Y. lumbered inside. "You should use a sleigh to get around today, veterinary."

"Not unless it's heated." Malcolm fished about for his chains and handles in the bucket but only found one set. "John, did you put your handle and chain in the bucket? I saw your son drop one in."

John Y. rubbed his beard. "No, veterinary, I handed them to you."

"I don't remember getting them."

"I think so"

Malcolm pondered the scene where he had labored. Straw bedding equaled lost equipment, especially in sub-zero weather. "Okay. If you find it forking out the pen, let me know."

Placing his hand on the milk house door handle, Malcolm hesitated, not eager to endure another long, cold ride. "Merry Christmas, again, John."

"*Danke, Tierarzt.* Merry Christmas as well. Are you sure you wouldn't want to take my sleigh home?"

"Come by and visit me at home later today if you would like to give me a ride."

"Perhaps after you take a nap, in case you didn't get enough sleep last night." John Y. winked at Jolene.

"Lord, this seat is as cold as that barn floor," Malcolm exclaimed, seating himself into Lucille. The side windows and much of the windshield were frosted. Malcolm huddled with Jolene under a blanket while the defroster cleared a small patch of visibility. Rubbing the glass, Malcolm hunched over to peek through. Jolene drained the thermos into the mugs, performing a Christmas miracle as the coffee was still hot. The snow had stopped, at least that which was traveling in a predominately vertical trajectory.

"God Almighty, a warm shower is going to feel good," Malcolm half-whispered.

"Me first," she replied.

"If that's the case, I'll wash your back."

She examined his soiled apparel. "If that's the case, you wash yourself first." They continued to slide and sway in silence as Lucille trekked across the tundra. Breaking the lull, she commented, "My presence will be a topic of conversation around Mr. Peachey's dinner table today."

"I would think the weather would be more worthy of discussion."

She smiled cryptically. "Not in Galloway County."

"That'll be a little weird next time I go to your folk's farm."

"Because?"

"Well, you know, because we're, um..."

"Screwing?"

"That's not *exactly* how I would have expressed that."

"How about screwing like bunnies on Christmas Eve?"

"Jeeez, Jolene."

"Don't be so uptight, they weren't born yesterday. Besides, I told Mom some time ago."

"You what?"

"We're pretty close." Malcolm slumped back against the seat. Patting his leg, she suggested, "Take off the harness, Malcolm."

"This is all very revealing."

"Biblically speaking, since you're divorced, we're committing adultery."

"You believe that?"

A corner of her mouth lifted slightly. "As should any proper self-righteous person."

"Biblically speaking, a woman who wears Levi jeans and cleans them on Sunday afternoons in an electric washer is an abomination onto the Lord."

"You forgot about eating pork for Sunday dinner." Jolene

scratched a small porthole in the passenger window. The swirling snow transported her back two years ago when she left Galloway County to indulge her adrenalin appetite for skiing in Vermont. A planned trip of two weeks turned into two years after meeting Kye. Her lips tightened. From one entanglement to another.

"You with me?"

She looked at Malcolm. "Remember how I told you that my fiancé, Kye, had been killed in a car wreck?"

"Does this snowstorm create an unpleasant reminder?"

"There's more to it. When they found him, he was in someone else's car, another *woman's* car. There was alcohol involved. She wasn't wearing any clothes but her fur coat."

"Jolene…"

She held up her hand. "He told me he was supposed to be at a meeting with the resort directors, down in New York. All along the bastard wasn't more than five miles away in this wealthy divorcee's winter palace. Turns out everyone at the resort knew they were an item for some time, all the while he's telling me he wants to get married. Needless to say, that has soured me on that institution."

"I'm sorry." He wrapped a blanket about her shoulders, "We'll be home soon."

Jolene looked at the ice forming on her side window, blurring reality in the white wilderness outside. She vowed not to lose herself in an illusory romantic attachment again.

CHAPTER XXXVII

The tempest of Siberian air swept from the valley with the same mad rush as when it arrived. Relentless cold and darkness continued in its wake, wearing on Malcolm's psyche. He toddled across the office, bottles and syringes bursting from his coverall pockets and arms crammed with boxes and vials. Pushing his back against the office storm door, he felt a tap on his shoulder.

Lois gauged the size of his haul. "Stocking up for the rest of winter?"

"No, last couple days were mighty busy, and John Y. Peachey called this morning for a drop off on the farm."

"Didn't want to stop by here?"

"Henry said it was okay. John Y. left to visit family for a couple of days. His sons are running the farm."

Lois's cleared the doorway to allow Malcolm to pass. "Might help to use a box, or make two trips, less prone to drop something."

"Good workout for the biceps."

As he unencumbered his arms onto Lucille's hood, her voice carried across the parking lot. "Dr. Malcolm, have you called Dr. Teesdale from the university yet?"

He grimaced. Unable to think of an excuse or make a quick exit, he responded in a flat voice. "No."

"Well, she's called the office two or three times now, I think it's high time to return her call." The tone reminded him of his mother drawing attention to an unsightly bedroom. "She said you can call her at home."

"What's the rush, Lois?"

Lois bustled into the office. "I'll get her number for you."

"How about just sending it over the radio?"

She poked her head back outside, cocked her head and scrunched her face. "How about I just get it for you?"

"Fine."

"Wait there," she sang out, "it won't take but a minute." Malcolm occupied himself by assigning bottles and vials into empty spaces and corners of Lucille. A finger inserted a piece of paper into his back pocket. "How about tonight then?"

"I'll see what I can do."

Lois withdrew. "Did you have a good Christmas weekend?"

"Rather on the cold side, Lois."

"I was thinking more of Christmas Eve, was that cold, too?" Malcolm's back spiraled at the waist, a box of mastitis tubes in his hand. She waved. "Have a good day, remember to call Dr. Teesdale."

The phone assignment weighed on Malcolm's mind. No doubt, the epidemiologist wanted to satisfy some academic curiosity, or subject him to a questionnaire that would 'only take 20 minutes of his time'. He disliked talking on the phone, especially when saturated from a string of night calls, it conflicted with his introverted nature. Oh well, better call. Lois will harass me until I do.

Malcolm managed to arrive at John Y. Peachey's an hour before evening milking time. He left the comfort of his seat and heating vents to gather up John Y.'s shopping list from Lucille. Malcolm shuffled through the snow in the barn yard and shoved against the heavy wooden door of the stable.

Cows lounged in straw-filled stalls, chewing cud in the sallow sunlight filtered by the frosted glass of the windows. Two small rat terriers appeared from a pile of straw in a back corner and sounded the alarm.

"You're alright," Malcolm consoled the sentries lowering his open hand. The dogs sniffed his hand, backed off, rumbled low growls and trotted away.

Malcolm searched for options to hang a note, preferably in an obvious place. He fixed his eyes on a clump of twine looped around a nail pounded into the far wall. Scuffing his boots between an alley of cow's tails, moans and rustles sounded to his right. Upon passing the last cow, he discovered an empty, well-bedded stall, empty from a bovine presence, anyway. Samuel Y. Peachey scrambled to put his feet into his pants, shirt unbuttoned, while a brown-eyed maiden held a blue dress over her otherwise naked body. Malcolm didn't recognize the girl but her face was as red as a tart cherry. He recoiled to his left, back-pedaling several steps.

"Sorry for the intrusion, Samuel. I was dropping off an order for your dad and I was looking for some twine to hang a note." Samuel Y.'s mottled face appeared over the back of a cow, his disheveled hair speckled with bits of yellow straw. Unsure whether he, or the young lovers, were the more embarrassed, Malcolm continued to step in reverse. "I'll just lay this with the medicine by the door." Malcolm all but fled the barn, surmising Samuel probably didn't know his dad had called the clinic for the delivery. He smirked. I guess Amish teenagers are no different than their English counterparts. Dale Calhoun had once revealed the parking hot spot for Amish buggies, an old gravel bank outside of Salty Dog.

Still trying to process his unexpected indiscretion, Malcolm found two calls waiting on his machine when he returned home. One was from Lois, reminding him to call the epidemiologist. God, she can be a pain in the ass! The other was from Jolene, with a proposal for New Year's Eve with Bobby and Marcy at the Dead Bear. She suggested such a fête would be educational, as he had never seen a New Year's Eve party at the Dead Bear. Malcolm fished out the note that Lois had dispensed into his pocket.

*****

Stephanie Teesdale, DVM, MPH, PhD, sat at her desk at home, reviewing computer printouts from the mainframe on campus. She had invested in one of the new personal computers, but it wasn't much help to her beyond word processing and spread sheets. She stroked her blonde hair, trying to decide on an experimental model for a clinical trial. The phone disturbed her musings. A glance at the wall clock revealed it was after eight. She picked up the receiver, answering in a composed voice, "Hello?"

"Hi, this is Malcolm Cromarty, a veterinarian up here in Galloway County. You've been trying to reach me about the rabies cases in our area."

Surprise and enthusiasm lifted her voice. "Yes, thanks *so much* for calling, Dr. Cromarty. I'm Stephanie Teesdale. I specialize in zoonotic diseases and have been calling your office this past week."

"I apologize for taking so long to call back. What exactly did you need to know about the cases?"

"Pretty much the history of each case. I already have the laboratory data on the breed, age, farm, and date of submission. I got that from Dr. Weinstock, the pathologist at the diagnostic laboratory."

"Yes, he confirmed the laboratory tests for us."

"Your county was in the middle of the raccoon-related epidemic this past summer. But I understand from your staff there were extenuating circumstances regarding the case in the ram."

She deemed there was a slight reluctance in his reply. "Well, there was a connection between the rabies case and other unusual local events."

"Unusual?" The pitch of her voice elevated again. "Listen, do you have a few minutes to review the case now? If not, we can arrange another time."

"No, it's okay. There aren't hard facts, only clinical intuition." He advised her. "It could take a while."

"As long as you don't mind questions along the way."

"No, that's alright."

She settled into her chair, scribbling notes on a pad while Malcolm recounted the experiences with John Y. Peachey's cow and the dog attack on Jonas P.'s sheep.

"So the neighbor's dog was never quarantined after the attack?"

"The owner wouldn't allow it."

"And it hasn't attacked the sheep anymore?"

"Not that I know of, I think that sinks my theory."

"Which is what?"

"The time span from the day of the attack to when the ewe died about two months later might be consistent with rabies, although I would have thought it to be shorter. But the slow progression of clinical signs in the ram bothers me. In contrast, John Y. Peachey's cow went downhill pretty fast."

"So what do you think explains all of this?"

"Given the aggressive nature of the black dog, perhaps it got tangled up with a rabid raccoon near the homestead. The dog wasn't ever vaccinated. Then it attacked the sheep and Mr. Yoder's farm dog, exposing them. How else do you explain the ram coming down with rabies and an exceptionally loyal border collie disappearing from the farm?"

"Is this black dog still alive?"

Malcolm sighed. "That's another issue with this line of thinking, I believe so."

Stephanie tapped her pen on the pad. This rural practitioner had clearly been contemplating the oddities of this case. "You realize that your sequence of events flies in the face of what we know of the epidemiology of rabies."

"I confess both Mr. MacDougall and his dog were partial towards unsocial behavior long before the day of the dog attack at Mr. Yoder's farm. And as you stated, we've had plenty of rabid wildlife, as well as a cow, in this county. I doubt those cases had any association with Mr. MacDougall and his dog."

She thought about his response. "Well, for starters, the time from the farm attack to development of clinical symptoms in the ram was, as you suggested, unusually long. You say Mr. Yoder lost a ewe earlier. Did he have any theories why it died?"

"He thought pneumonia, but I doubt he knew the true cause."

"And how could Mr. MacDougall's dog be alive all this time if it has rabies?"

"I guess that will be one of life's mysteries."

She held her breath and offered a comment that went against her scientific grain. "Perhaps, but what intrigues me is that there may be some parts in your line of thought that are plausible." She paused for Malcolm to react, but he remained silent. She continued, "When I looked at the data from all the cases reported from your and surrounding counties this past year, no matter what type of animal was submitted, they were infected with a variant of rabies virus that originated from raccoons."

"That can be determined?"

His diagnostic *naiveté* amused her. "Yes, from monoclonal antibody testing. But here's the catch, the rabies virus variant isolated from Mr. Yoder's ram originated from a bat."

"Could this variant allow an infected animal to harbor and transmit the virus longer than usual?"

"That's an exciting and disturbing scientific possibility, isn't it? The bat that infected this ram may have been a carrier for a strain of virus with atypical qualities! Are there any caves in the area?"

"More than a few, although I don't know of any personally. And there are lots of old barns and outbuildings throughout the valley, some in various stages of disrepair."

"Bats are hibernating or have migrated to milder climates this time of year. I would like to visit Galloway County in warmer weather, when they are more active. Does Mr. MacDougall live close to Mr. Yoder's place?"

"Unfortunately, for that particular Mr. Yoder, only a half mile away."

"Excellent, so we have compatible geographic distribution. Well, perhaps we can collect and test bats for the virus when spring returns."

"I'll see what I can find out about caves up on this end of the valley, nearer to the MacDougall place."

"That would be great. You are an acquaintance of Mr. MacDougall as well?"

There was a brief lull. "Not too close, no."

"Thank you so much for taking the time to fill me in on this,

I'll be in touch when spring returns."

"No problem."

"Until later." Stephanie hung up. This would be the kind of groundbreaking discovery that would propel her into gaining grant funding and tenure. She needed to get a research team together. A state road map rested in a lower desk drawer, time to determine the best route to Galloway County.

CHAPTER XXXVIII

Disconnected from Jolene in a blue haze of cigarette smoke, Malcolm bumped along like a pin ball, tracking towards the bar. Gaudy silver and gold decorations festooned the walls and mirrors behind Katy and another barmaid he hadn't seen before. Their nimble hands were hard pressed to fill glasses and dispense beer bottles to the ever-thirsty mob.

A beefy paw slapped Malcolm's shoulder. "Follow me Doc, we have plenty of drinks over here." Dozer cleared a corridor through the densely clustered patrons.

"A little lost?" Bobby yelled over the din.

Shannon, a beer in each hand and flanked by Jolene and Marcy, greeted, "Happy New Year, Doc."

Ronnie stood by his side, his silver *lamé* bowling shirt shimmering under a black tux jacket and a maroon fez cap.

Shannon leaned closer. "Some party, eh Doc?"

Malcolm registered a one-hundred eighty degree view. "Got to admit, it's one of the most unusual gatherings I've ever witnessed. Is this a typical New Year's here?"

"As long as the management continues to allow it."

A scuffle on the far side of the bar caught Shannon's attention. He shook his head, "Always has to be somebody that forgets their manners." Dozer plowed through the crowd, homing in on the altercation.

Malcolm surmised that the giant's mission wouldn't bode well for the troublemakers. "Is that difficult to do?"

"How's that, Doc?"

"You getting along with management?"

"Naw, we're in with them real good, aren't we Jimmie?" Jimmie managed an uneven smile.

"So are we going to try a little dancing?" Marcy prodded Bobby.

Malcolm asked, "What style of dancing is that, exactly?"

"No style much involved," Shannon replied. "Take your own swing at the plate, so you do."

Ronnie said, "And let the mood take you." The fez tassel started spinning.

"Take the likes of you where?" Shannon yelled.

"What?"

"I said you ain't much of a dancer."

"I can cut the rug about your goddam clumsy feet!" Ronnie retorted.

"That's not cuttin' a rug, that's smashing cockroaches."

"I'll get out there and show you," Ronnie persisted.

"You and what poor unwitting partner?"

"Why me of course," Jolene said, stunning the discussion into silence.

"Ha," Ronnie ejected, "By the end of the evening, all the ladies will want to follow suit."

Malcolm took note of Ronnie's feet, penny loafers—a complete outfit indeed.

The band ceased their set for a break, decreasing the decibel level to casual bedlam. Malcolm gathered his thoughts. "There is something of interest to me that valley natives such as yourselves may know about." Jolene abruptly halted sipping her bottle. "I chatted with a public health specialist from the state university. She's curious about the rabies outbreak here because of an unusual type of rabies virus originating from bats."

Ronnie grabbed Malcolm's forearm. "You mean to say we've got rabid bats flying around in the valley?"

"Certainly not during winter, and I doubt they're flying around looking to bite people."

Shannon puckered his lips and whistled. "I'll be damned."

"Bats are a menace, always said that." Ronnie looked towards the ceiling.

"They're no more a problem, maybe less so than any other wild animal and I personally admire their appetite for flying insects," Malcolm said. "What I'm trying to find out is... are there any caves in the valley, especially near the upper end of the valley?"

Jolene folded her arms. "Don't you also mean near John MacDougall's place, Malcolm?"

"Well, I guess."

The Dead Bear elders put their heads together. Shannon spoke first. "Bunches of places like that up in the hills, especially in some of the spurs along the mountains. One of the bigger ones is Dark Spring Cave."

Ronnie said, "Caves under this valley are linked by an underground maze. If you're clever enough, you could go from one end of the county to the other."

"What kind of horseshit is that?" Shannon blurted.

"What about Enos M. Byler's tractor that fell into that sinkhole a few years back?"

"That doesn't make a whole underground tunnel system."

Malcolm asked, "So, you think this Dark Spring Cave, or others in that area of the valley, might have bats? The university investigator wants to trap some when the weather warms up."

Dozer reappeared, several beer bottles clasped in his hands. Malcolm turned to Bobby. "Who's got the bar tab?"

Bobby said, "I wouldn't worry about it. Owners are covering for us."

Malcolm scanned the masses. "Are they here?"

"Man, you don't know? Take a wild guess." Malcolm followed the trail of Bobby's attention towards Shannon.

"It couldn't be."

"Shannon, and his partner."

"Not Ronnie!" Jolene exclaimed.

"Jimmie. Jimmie and Shannon."

The band was staggering back to their instruments, well lubed after spending their break time over a beer.

"Let's go cool down those hot feet of yours," Marcy cradled Bobby's arm.

Jolene clasped Ronnie's hand. "Ready Mr. Rug Cutter?"

Malcolm's sight snaked through the crowd with Jolene, a goddess among mortals. Shannon broke Malcolm's concentration. "So I gather your gal doesn't take kindly to your fretting over MacDougall and his dog again, Doc."

Malcolm replied, "Odd that I haven't heard much about that beast lately. Have you?"

Shannon scratched his head. "No, can't say I have. The damn thing must be holed up for the winter."

Malcolm studied Shannon's face. "Perhaps."

"Heard any more about the missing pets?"

Malcolm reflected on his visit from Cal a week earlier. "Only what I see on the posters around town."

Shannon pointed towards the door. "Still on the bulletin board here, too. It's a little peculiar having so many missing at one time."

The music ramped up and Shannon increased his volume to a shout. "Them caves up in the ridges were handy places for the moonshine business, back in the day."

Malcolm took a swig from his bottle. "More lucrative than farming I guess."

"I told you before of ol' Jeb MacDougall and his boy Trotter running moonshine around here. And that Trotter was quite the hell-raiser."

The old photos that Jolene had researched crossed Malcolm's mind. "So I've heard. Apparently he got into robbing banks and shoot outs with the police."

"You've got that right, Doc."

"What on Earth made him do that?"

"Dutchmen."

"You mean the Amish farmers?"

"You see, Jeb had to sell most of his tillable land to a dutchman to raise cash during the depression, after prohibition was repealed and took away their money from the still. That was bad for business." Shannon shook his head. Jonas P. Yoder's folks, I think it was that bought the land."

"Cause for a little resentment?"

"Simmered for a few years after the sale, then boiled over into downright spite. One fall day, Trotter went into town, three sheets to the wind and sees the banker that almost foreclosed the MacDougall place. As it was, Trotter took to yelling at the guy about dutchmen and bankers. The banker tried to pay no mind, which torqued Trotter all the more. He knocked the banker to the ground, looking for a fight. Couldn't get one, so he pulls out a gun and forces the poor man back into the bank and demanded the return of the family fortune."

"Did he shoot the banker?"

Shannon shook his head. "Not him. One of the tellers set off the alarm, the township cop showed up and when he called out for Trotter to quit, Trotter shot and killed him right then and there with just one shot. Trotter grabbed some cash and took off in his souped-up Ford coupe that he used for boot legging."

"This all happened at the old state bank in the center of town?"

"Yupper. After that, Trotter drove down to Fultonville to rob another bank, and on the same day, two banks in Gillstown. No shame, parking his car and going in like he was asking for a business loan. He ended up back at the MacDougall place. State troopers chased him up there and got in an awful shootout with Trotter, Jeb and their hired man; killed all three of them."

"Good God!"

"Trotter was shot to pieces right by the bumper of his Ford."

"Yeah, I already got wind of that piece of information."

"Crazy thing of it is that clan made money hand in fist on liquor, loads more than they ever did on that sorry farmstead. But Trotter couldn't let the land sale be, got fixated on it, so he did."

"All this over a land sale? You can't be serious."

"Serious as a heart attack, Doc. By the way, you know the township cop that was killed at the bank? He was Dale and Jack

Calhoun's uncle."

Malcolm felt that he was no longer sliding on a thin sheet of ice but had crashed through into deep dark waters.

Jolene returned with Ronnie, who was out of breath, but beaming ear to ear. She said, "I'm all warmed up and ready to go, are you?"

Malcolm regarded her with a momentary lapse of recognition. He put down his beer, distractedly answering, "Sure, I mean, yes, let's go."

"What's the matter, afraid you can't keep up with Ronnie's pace?"

A brunette in a tight blouse hooked Shannon's arm. His blue eyes feigned surprise. "Well, time for this rooster to prance about the barnyard with the hens."

Malcolm asked, "Say, Shannon, you wouldn't know who owns a big chocolate lab? A big *male* chocolate lab?"

The barber and beer merchant of Salty Dog wheezed out a laugh. "Why Doc, you always seem to be searching for the owner of one dog or another."

"Come on, time to dance." Jolene objected. Once again, Malcolm lost track of time and place, succumbing to the spell of emerald gems within a cyclone of auburn hair.

Auld Lang Syne, caterwauled by a hundred voices, ushered in the New Year, followed by three more choruses, without regard for matching lyrics.

Near closing time, Malcolm, Jolene, Bobby and Marcy collected themselves to leave the party. In the piercing cold air of the parking lot, a figure in a grimy coat and brimmed hat intercepted them. Malcolm looked into the beady eyes of Haumiller with distaste.

"Out for a party with the little ladies, Mr. Vet?"

The alcohol in Malcolm's veins responded gruffly, "Take off!"

"Oh, Mr. Hot Shot Vet, shouldn't throw stones if you live in a glass house." Haumiller sneered, "That is, houses with glass windows with mouthy dogs."

Malcolm lunged at Haumiller, sizing up his chin with his

clenched fist. Bobby latched onto Malcolm's shoulder while Jolene blocked his assault. She pleaded, "Let it go. You're better than that." Bobby tightened his grip and Malcolm relaxed his hand.

Haumiller cackled, "Yeah, glass windows in houses like yours."

With the fire of dragon's eyes inches from his face, Jolene seethed. "Back off, or I'll kick your ass myself."

Haumiller flinched, wavered backwards and staggered away.

Marcy asked, "Who was that?"

Malcolm watched the hat dissipate into obscurity. "Someone who doesn't like the way I handle emergency calls."

Bobby said, "He probably had nothing to do with the windows in your house. Everyone around here knows about the shooting. Just flapping his drunk-ass mouth."

Jolene's hand replaced Bobby's on Malcolm's arm. "Let's go home."

Precious didn't meet them at the door. Tapping the light switch, Malcolm called out, "Precious, you here?" He paced through the rooms. "Precious!" Cold spasms seized his back as he called her name again.

Laughter peeled from the bedroom. Malcolm burst into the room and found Precious lying in his laundry pile in the closet. He knelt down and stroked her head. "What have you gotten yourself into, pretty girl?" Three brown puppies squirmed by her side.

Jolene said, "I guess this makes you a godfather of sorts."

Malcolm muttered, "Yeah, and from the looks of them, no doubt who their sire is."

CHAPTER XXXIX

Come hell or high water, Chief Cal was going to train his nephew how to cope with problems other than drunk driving and speeding violations. By Cal's reckoning, Tommy wasn't the brightest bulb on the family Christmas tree and the wires in his brain tended to get tangled when connecting observations with deductions. The young man was steady enough, didn't get flustered in tough situations and with a rifle in his hands, was a sure bet to take a buck at three-hundred yards. Parking his car by the front porch of the homestead, Cal turned to his protégé. He jerked his thumb towards himself and said, "Now listen, let me do the talking and keep your eyes open."

Cal zipped up his coat and folded the collar to cover his neck as he got out of the car. The late afternoon sun smoldered through a muslin sheet of thin clouds above the ridge to the southwest, doing little to keep the breeze from biting his ears and nose. Cal steadied himself and pushed up the steps slowly, while Tommy remained by the front of the car. Staccato echoes of knuckles rapping the door ricocheted throughout the clearing.

A hand dragged the door open and John MacDougall leered through the screen door, wearing torn jeans and a tee shirt blotted with a brown, oily, paste. His scraggly beard and unkempt hair looked as if a washing was overdue for at least a fortnight. Despite the lack of hygiene, like a cheap cologne, the scent of whiskey masked any odors from the soiled clothes.

MacDougall stared at Chief Cal through heavy eyelids. "Something you want?"

Cal caught a glimpse of what seemed to be the same paste that adorned MacDougall's shirt smeared in his hair. "Morning, John. Thought we'd stop by to see how things were going."

"Well enough." MacDougall's eyes flashed beyond Cal's shoulder in Tommy's direction. "Who's that?"

"Why you know him, John. That's my nephew Tommy…mind if I ask you a couple of questions?"

MacDougall leaned heavily against the door jamb. "Like what?"

"John, it seems that some pets have gone missing in the valley."

"So what?"

"Haven't seen anyone unusual about?"

MacDougall jutted his jaw. "I know what those bastard Calhoun boys and the fucking dutchmen are thinking, Archie's to blame. He ain't around, you hear? He had nothin' to do with what folks said he did back then. And he sure ain't doing nothin' now."

"Now don't get riled up, John. I said nothing about Archie. I'm only trying to piece things together."

"Archie got all fucked up while those damn dutchmen sat on their sorry, yellow, chicken-shit asses back here."

Cal responded in an even tone. "John, I'm sure your brother served his country well. But I have to ask if folks in the area have seen any suspicious people or things going on."

"Hell no, I haven't and Archie ain't here!"

Cal rubbed his neck. "That's fine, John. Thank you for your time, we'll be on our way."

Cal turned about and lowered himself down the steps. When he reached his car, he turned to MacDougall. "You will let me know if anything out of the ordinary happens?"

MacDougall reset his hold on the doorframe. "Tell them dutchmen to keep away from here!"

Cal glanced at the tree with the chain wrapped around the trunk, the snow underneath was undisturbed. He surveyed the yard for tracks—none were to be seen. "Oh by the way, John, how's that dog of yours doing?"

"Don't know, he goes where he pleases."

Cal gestured to Tommy to get back in the car and eased himself behind the wheel.

Tommy folded his arms, rocking in his seat as the Ford plowed in and out of the ruts in the lane. "MacDougall's been drinking, no doubt."

"No doubt."

"Who does he think he is, sassing back at the law like that?"

Cal turned onto the road. "He's troubled, all right. But he keeps to himself and would rather if other folks did the same." Cal dialed the heat fan up to high. His fingers appreciated the cold little better than his knee.

"But Dale Calhoun says…"

Cal clenched the steering wheel. "Dale Calhoun be damned! The rear brakes squeaked as the Ford swayed around a curve.

"We could sure use some new shocks, Uncle Cal."

"Tell that to the village supervisors."

"Kind of fun though, bouncing around like a four-wheeled pogo stick."

Cal arched his brow. *Lord, give me strength.*

*****

John MacDougall pushed his glasses onto the bridge of his nose, in time to see the police cruiser leave his property. Of all times for Archie to reappear. Fuck Cal and his nosy-ass self and his stooge of a nephew. Fuck the Calhouns. Fuck the dutchmen and their goddam sheep. Fuck that vet, fuck people and their missing pets. He leaned on a worn oak chair and reached for whiskey. How was he going to keep those meddlers from messing with his brother?

His bloodshot eyes settled on the kitchen counter. The peanut butter was near gone and so was the bread. Pieces of bacon lay sprinkled about the kitchen, remnants of his sandwich making frenzy. Time to go to the market to stock up again, Archie had an appetite.

CHAPTER XL

The gears of winter were stuck in neutral, jammed in snow-packed barnyards, roads, and farm lanes. Lucille probed through three inches of fresh snow between the ancient locust trees along Jonas P. Yoder's lane, their bare limbs clutching at the dismal clouds above them. The yellow light of the stable brightened an otherwise anemic landscape.

The farmer and his oldest son squatted on their stools, squeezing milk into pails. Jonas P. directed Malcolm to a box stall in the back of the barn. Malcolm treated his recumbent patient by flashlight and without incident. Returning to the stable, Malcolm found Jonas P. hefting two pails brimming with the smell of butter fat to the milk house. "So, *Tierarzt* how is the *neu Jahr* for you?"

Malcolm shrugged. "No complaints." He tagged along in the farmer's footsteps towards the milk house. A young boy was scrubbing a calf bottle in the sink.

"*Morgen*," Malcolm greeted.

The boy looked up and smiled, Malcolm recognized him as the covert Bugs Bunny fan from Dale Calhoun's. Jonas P. hoisted the pails and poured the warm milk into ten gallon cans set in the spring fed water tank. "Busy day today, *Tierarzt?*"

"I haven't been in the office yet to find out." The boy retrieved one of the pails from his father to wash in the sink. Malcolm watched his actions. Farm kids do this every day before they even get to school. He inquired, "Do you like school?"

The boy lowered his head and smiled. Jonas P. stroked the back of his son's head. "Sammy does well in school. He's good with numbers.”

“Which schoolhouse? The one up by Enos M. Byler’s harness shop?”

“*Ja,* not very far.”

Malcolm dwelled on Jonas P.’s definition of ‘not very far.’ It was an honest mile. “I imagine he and his schoolmates haven’t seen much of the black dog during the winter?”

“I don’t think so. Eh, Sammy, *hast du den schwarzen Hund gesehen?*”

Sammy snuck a furtive glance at Malcolm. Jonas P. leaned over his son. “Sammy, *hast du ihn gesehen?*”

Sammy squirmed and mumbled, “*Er ist tot.*”

“*Was?*” Jonas P. exclaimed.

“*Er ist tot,* daddy.”

Jonas P.’s lips trembled. Malcolm darted his eyes between Jonas P. and his son. “Did I understand him to say ‘dead’?”

Jonas P. snapped, “*Wo war das? Wann?*”

“*Im Herbst. Am Straßenrand. Seth war mit mir. Ein Tag, als wir zur Schule gingen.*”

“*Stimmt das auch?*”

Sammy’s voice faltered. “Ja, daddy. *Ich sah das Blut.*”

Malcolm lifted his hands. “What did he say? Did he say back in the fall?”

Jonas P. placed his hands on his hips and scrutinized the boy. “*Ja,* by the side of the road, on the way to school with his brother, Seth. He saw blood.”

Malcolm put his hands on his knees to gain level eye contact with Sammy. “Did you go near the dog?” Sammy didn’t speak.

“Sammy!” Jonas P. commanded, “*Gingst du nahe dem Hund?*”

“*Nein.*”

Malcolm let out a long breath. “Sammy, this is good to know, *danke.*” As he washed his boots, he tilted his head to the farmer to step outside. The faint blush of dawn was strengthening. Malcolm said, “Sammy hasn’t done anything wrong, he’s just afraid. Ask him gently to make sure he didn’t touch the dog. Make sure Seth didn’t get near it either.”

Jonas P. regarded the milk house. "They should know better to tell me such things."

"Maybe they didn't know what to make of it."

Malcolm finished his thought as Lucille scrambled up the farm lane through white powder sifting in the air…or maybe they saw something they shouldn't have.

*****

Malcolm blundered into the office and a taut discourse.

"Shouldn't be surprised they would react like that."

Henry, sitting on a bench and sipping coffee responded, "No, Paul, I suppose not." Lois tapped a black pen onto a type-written letter resting on the desk.

Malcolm surveyed the trio and informed them. "Guess what? Jonas P. Yoder's sons saw MacDougall's dog, dead on a roadside." His audience paid no heed. "You know…the big black dog that attacked Jonas P.'s sheep?"

Henry said, "A good end for that public nuisance."

"Yeah, but what if it had rabies? Now we'll never know."

Rubbing the back of his neck with his large, thick fingers, Paul responded, "That's insignificant right now. We have a more demanding issue on our hands."

Lois swept the letter into her hand and presented it to Malcolm. "This came in yesterday. It'll be hitting the news soon enough, if not already."

Malcolm scanned the state government letterhead and the key phrases of rabies, dog vaccination law, and a deadline for compliance. He pitched the letter back onto the desk. "I think it's long overdue."

"I don't doubt that, Doctor." Paul's voice had an acerbic accent. "But given that we have only six weeks to get every dog in the state vaccinated, this seems a little extreme."

"Six weeks?" Malcolm re-examined the letter. "Six weeks?"

Henry added quietly, "As I understand it, the fine for non-

232

compliance increases as time after the deadline increases."

Malcolm calculated the math. "That means we might have to complete this by about St. Patrick's Day."

"That would be six weeks," Paul added.

Henry stroked his whiskers. After a few moments he concluded, "Paul has been doing the small animal work for us two days a week since his retirement. I think we'll have to offer vaccination clinics on a third day, and probably a couple of evenings a week.

Drumming her polished red nails on the desktop, Lois suggested, "Someone could possibly handle several hundred animals in an afternoon if we set up a vaccination clinic in a public place like a fire hall." All heads turned her way. "I'll bet you could get lots of volunteers. Let them fill out the paperwork, organize the lines, and hold animals; you could line up and shoot. If we got three bucks a shot, a couple bucks for the volunteer fire company, and two more to cover the vaccine and supplies, everyone would come out ahead, and still be affordable for folks who don't have a lot of money."

Henry studied the coffee mug in his hands.

"Henry, it would be a good bit of civic outreach, something the whole community could get involved in."

Henry granted a tentative blessing. "It might work."

"Hold the clinic on a Saturday," Lois looked pointedly at Malcolm. "Perhaps a young, eager veterinarian on his day off." Then she bore her eyes into Henry's. "Perhaps pay a bonus for his time."

Malcolm regretted his previous bravado. "Well, I guess I did say mandatory rabies vaccinations were overdue."

"And you *personally* know some well-connected people to find some help", Lois added.

Malcolm yielded. "I guess I could pull this off. It will be an interesting experience."

Henry nodded. "Ancient Chinese curse, may you live in interesting times."

"Speaking of pet practice, I need to get in some day and spay Precious."

Paul lamented. "Now? With everything going on?"

"Well, she'll be weaning her puppies soon."

"Puppies! You didn't spay your dog until she had puppies?"

"I got her from a shelter; I was told she was spayed."

Paul threw his hands up and tramped to the back room. Lois smiled at Malcolm. "What kind of puppies?"

"Mostly lab, like their dad."

"You know who the dad is?"

"Broke into my yard; don't know the owner, though."

Lois said, "Henry, you could use some company inside that stark old farmhouse of yours."

Henry stretched up from the bench and directed himself to the back room to rinse his coffee cup. "I don't know if I need that kind of responsibility right now, Lois."

With Henry out of earshot, she whispered, "I'll get him to take one. He needs a friend."

Malcolm cocked his head. "How do *you* know about the interior of his farmhouse?"

Lois' eyes kindled from within.

CHAPTER XLI

The door flung open as Dale and Jack Calhoun erupted into the police office. Next to their side was a stout, middle-aged man in an orange vest and cap, carrying a feed bag that was weighted at the bottom. Cal recognized him as Louis Tallman, kin of the Calhoun's.

Speculating as to the contents of the bag, Cal greeted, "Good morning, boys."

"Good morning my ass," Dale snorted. He grabbed the bundle from Louis, approached Cal's desk and flipped the sack over. The body of a black and white cat flopped out; blood smearing the papers on the desk from the stump of a missing tail.

Cal leapt out of his chair, balancing himself from tripping on his bad leg. "What in blazing hell do you think you're doing?" He glared at Dale, "Get that goddam thing off my desk!"

"Well, you keep that goddam bastard from killing folk's pets, or I'll shoot the son of a bitch!"

Scowling at Jack and Louis, then back at Dale, Cal spoke through gritted teeth. "I said get that thing off my desk!" Tommy appeared in the doorway that led to the jail cells.

"It's time you did your damn job, Chief. That son of a bitch is causing a fuss, might be his brother is back." Dale pointed to the carcass. "Louis's mother found her poor kitty this morning on her front yard. She's nearly 80 years old and has to see this!"

Cal appraised the cat. The eyes were a pair of dark stones that reflected no light. He rubbed his brow. "Dale, that could have been coyotes."

Jack sneered, "Coyotes?"

Dale added, "Don't you think we ain't figured out who shot our cow a while back? You know who it is."

Cal grasped the carcass. "Open that bag!" Dale faltered. Cal roared, "Open that goddam bag, or I'll bust you for assault!" Tommy unhooked his holster.

Dale blinked and pulled the edges of the bag. His arms sank as the remains dropped into the bottom of the sack.

"Now listen up, boys. Listen up good. I'm looking into this, but I warn you not to do anything that you'll regret, you hear me? I'm sick and damn tired of this whole thing between you and MacDougall. You've gone and dragged that young vet into your high jinks and he's way out of his league." Cal glowered at Dale. "If anything happens to him, or MacDougall, or anyone else, I'll nail your sorry hides inside that jail and let you hang there until you rot." He slowly turned his head to Jack. "Kind of like visiting old digs for some of you."

The faces of the trio were stoic as carved granite. Dale murmured, "If I see anyone on my property that don't belong, I'll shoot first, then ask questions."

"I'll haul your ass in for manslaughter."

"I'll just be aiming at a coyote."

"Get out!"

The war party retreated, Jack lowering his head in an ill attempt to mask an ugly grin. He muttered under his breath, "Coyotes, my ass."

Cal watched every step of their progress and let out a breath as the door slammed. "Tommy, please get me a couple of paper towels rinsed in cold water."

CHAPTER XLII

Drop by drop, Malcolm witnessed winter dissolve into spring from an icicle over his kitchen window. Even over the phone, he could tell Lois needed a favor.

"We got a call from someone who is hard pressed to get to the fire house next Saturday." Malcolm closed his eyes and twisted the phone coil. "An elderly woman with eight or nine Pomeranians. She can't get them all in a car."

He sighed. "Imagine that. Stuffing that many hyperactive dogs into a vehicle is a job better suited for circus clowns."

"Shame on you, she's elderly!"

"She wants me to come to her house, Lois? On the day of the vaccination clinic?"

"Yes, that's what she asked."

Jolene padded softly into the kitchen, cradling a chubby puppy in her arms. Malcolm rubbed a finger along the plush spine. "I should be organizing the fire hall for the clinic that morning."

"The firehouse volunteers can handle that. You can't visit the following weekend because you'll be on call and with the thawing weather; you know we don't have time during the week because we're so busy with calf work. Surely you can find someone to help speed things up so you can make sure you reach the clinic on time?"

Malcolm glanced at Jolene. "Where does she live?"

"West of Gillstown on the river highway."

"That will take thirty minutes to get there, probably an hour for the shots, and another half hour to the fire hall."

"That's two hours. Let's throw in some prep time at the clinic, how about you plan to meet her at her home at about nine?"

"Swell."

"You're a peach. By the way, I'd take all of the rabies vaccine we have in stock for your clinic next Saturday."

Malcolm dismissed her. "We're not going to have *that* many."

"You never know. Take all of them."

"As you wish." Malcolm slapped the phone into the receiver. "My surrogate mother!"

Jolene deposited her snoozing ward into his hands. "And here's your surrogate child."

"He's a plump little rascal, isn't he?"

"I think he'd make a good companion for Delmar."

Malcolm's eyes widened. "What of Pierre?"

"Pierre is my dog and Delmar needs some company."

"Pierre is enough company for ten people. "Did you check this out with Delmar?"

"Doesn't matter, Mom thinks it's a good idea."

"I imagine the puppy will have to listen to all of Delmar's colorful UFO theories."

"Now you know why Mom thinks it's a good idea, time for a new pair of ears for the lectures."

*****

The next weekend, Malcolm navigated his Scout into the Salty Dog town square.

Jolene gasped, "What in the hell?"

Countless vehicles choked the narrow streets into a virtual standstill. Jolene spun her head in all directions as scores of dogs and their owners filtered around them. "Geez, there must be a couple hundred lined up in front of the fire hall alone."

A gaunt, tall man galloped past Lucille desperately holding onto the leash pulled by a cavorting Newfoundland. Frustrated owners tugged on pets that dug their paws in the muddy ground and extended their noses to other noses and rear ends. Soulful captives hid in pet carriers, appearing cold, miserable, and anxious.

Jolene exclaimed, "Good Lord, I think I saw a couple of goats."

Malcolm trickled his Scout through the bedlam, trying to reach the center of activity next to the fire house. A volunteer fireman recognized Malcolm and directed him to a safe landing in front of the building.

Malcolm estimated the numbers in the queue of waiting patients. "This is out of hand." His posture stiffened. "There he is, the bastard!"

"Who?"

Malcolm pointed across the road. "There's the chocolate lab that bred Precious!"

"Are you sure?"

The canine philanderer was free of any restraint and created mayhem and envy among a fair portion of his peers. Malcolm pulled his door handle. "I'll fix him."

"You don't have time to chase him around town now."

The dog spied Malcolm, strutted over to a fire hydrant, lifted a leg and withdrew down the street.

"Cheeky bastard."

"Yeah, yeah, let's go Dr. Doolittle."

Tables stretched end to end halfway across the meeting hall. The smell of popcorn and hotdogs wafted from behind a window counter laid out with bottles of ketchup, mustard and relish. A clown in a red wig and yellow jump suit pressed green and yellow balloons onto a helium tank. As one of the balloons escaped up to the ceiling, the clown spewed a reedy cackle to the extent that Malcolm questioned his sobriety.

Malcolm scanned the layout of the improvised clinic. "I didn't realize this would be such a carnival."

Jolene nodded in the direction of the other side of the hall by the far door. Shannon, Jimmie and Ronnie, the latter in a red and black bowling shirt, manned a table set on a dais adorned with red, white, and blue bunting. An electronic Bingo score board hung behind them on the wall. "Looks like you have all sorts of help."

Shannon beckoned Malcolm with a wave. A pitcher of orange

juice, plastic cups and a cooler rested in front of him. "How do you like our judge's table, Doc?"

"Judging for what, Shannon?"

"We're selling entry fees so folks can participate in the first annual Salty Dog - Dog Show."

Malcolm folded his arms. "Dog show?"

"We'll select, after careful deliberation, the best looking, and" —he looked about and lowered his voice—"the butt-ugliest dog in the show. We'll have monetary prizes for both the winners."

"Will someone really want their dog to win the ugly award, Shannon?"

Shannon winked. "Don't you worry none, Doc. We'll call it the best-matched-to-their-owner award." Shannon called out, "Jimmie, would you kindly get Doc and Jolene some cups of juice?" Jimmie set about filling cups, including refills for Shannon, Ronnie, and himself.

Stepping down to the floor to serve the beverages, Shannon placed his hand on Malcolm's shoulder and drew his attention to the tables on the other side of the garage. "You see, Doc, you'll give the animals their rabies shot over there. Then they walk by in front of us on their way out. If they have an entry ticket, they can tell us their number and we'll write down a score for the dog. Five dollars a ticket, eight will get you two, if you got two dogs. We're selling them at the concession stand, while folks are getting themselves a hot dog."

Malcolm pondered. Nothing says professional workplace like hot dogs and clowns.

Jolene inquired, "And you guys are the judges?"

"Three of them. Then there's ol' Ralph Spicher, he don't see so good, but he's the president of the fire company. And that there's Agnes Hargrove, she raises purebred collies and is the chairwoman of the Galloway County Kennel Club."

"She doesn't raise collies, it's Golden Retrievers," Ronnie said.

Recalling their previous expertise on canine pedigrees at the Dead Bear, Malcolm figured there wasn't a more qualified panel of judges in all the land. "And the two other judges know of the 'ugly' category?"

Shannon's eyes narrowed. "Ralph won't be able to tell a Greyhound from a Chihuahua and who cares what Agnes picks? She's a bit full of herself, head of the local library board, DAR, lady's church league, and takes to gossiping with a vengeance."

Malcolm thought. Present company as well.

Jolene was grinning. Malcolm took note and sampled his beverage. Carbonation tickled his tongue and throat. "Shannon, what is this?"

"Why, Jimmie's made up a batch of mimosas, Doc." Shannon lifted his cup to a bag of chocolate doughnuts on the table. "They'll go great with our brunch."

Malcolm withdrew to the working end of the clinic, abandoning Jolene to the contest. Sally Calhoun sought him out to give him instructions. He only had to examine the pets, administer vaccinations and sign certificates. The fire company volunteers would handle everything else.

A middle-aged man and his spaniel stood first in line. "Doctor, Herb Getz here. It's a fine thing youens are doing today, been in line since 8:15 AM."

"You've been waiting for four hours?" Malcolm rubbed the backside of the dog.

"It's a fine patriotic thing to do," Mr. Getz continued.

"Pardon me?" Malcolm asked abstractedly as he injected the vaccine.

"This here rabies outbreak, all started by communist agents and other infiltrators. Infected our wildlife so they did."

Malcolm's jaw dropped. "I'm sorry?"

"Yes sir, we must be ever wary of their schemes. They use satellites and send signals to their spies in this county, even as we speak."

Without looking at the certificate, Malcolm aimlessly scrawled his signature. "Here, in Galloway County?"

He leaned closer to Malcolm. "And UFO's? Why they're

nothing more than special reconnaissance for the ..."

Sally Calhoun rescued Malcolm. "That's fine Herb, but we have to keep things moving today."

Mr. Getz called over his shoulder. "I got books about how they pushed fluoride into our water to sterilize us. You can borrow them anytime, doctor."

Shannon called out, "Don't you want to buy a ticket to enter the contest, Herb?"

"I'm not spending one extra dime on any fire company that's going to spray fluoride-tainted water all over my house to put out a fire," Mr. Getz retorted. Shannon guffawed. Jolene sat next to him with a pad of paper, pencil, and a cup of refreshment. She eyed Herb as he made for the exit and gave Malcolm the 'thumbs up'.

Malcolm let out a long breath and set about the task of giving shots and signing papers. Try as he might to resist temptation, he found himself evaluating the dogs for physical appearance, crafting his own choice for the most homely.

The afternoon wore on and people and their pets kept flowing by in a steady stream. Malcolm took a quick break. Devouring a hot dog, he neared the judge's table and nodded hello to Mrs. Hargrove and Ralph Spicher. The latter's eyes twitched between snooze and remote awareness, struggling to fan off a siesta. "How's it going over here?"

"We've had over ninety contestants lay down their entry fee, Doc." Shannon said.

"How can you keep all the dogs straight in your mind?"

Shannon whispered, "Truth be known, it's kind of hard to recall, but we're still keeping our eye out for that special couple. We've changed the class to the ugliest dog and worst dressed owner combined. It was Jolene's idea."

"It's for a good cause," she added. "We all think you look very professional over there."

"That's a great consolation."

Mid-afternoon dragged well into the evening before the deluge of enervated pet owners and their charges ebbed into a trickle. Lead replaced marrow in Malcolm's bones. Shannon crowed that the winner of the first annual dog show was half German Shepherd, half Dachshund. Malcolm was too weary to extract an image of the critter from the blur in his mind.

Sally said, "Dale wanted to come, but he needed to help Jack fix a hydraulic line on the tractor. Thinks he can plow within a week or so if it don't rain too much. You know farmers, always fixing their toys in the machine shop."

"Not to worry, we did fine. Tell him hello."

Jolene came by. "Hey, Sally, how are you doing?"

"Well, thank-you, how did youens do at the judging table?"

"Great, after the prizes, there must be nine hundred bucks for the fire company."

"Mercy! Plus the share from all those shots, and concessions, we should do one every year."

Malcolm's insides churned.

Shannon wandered over, "The boys think you deserve a beer at the Dead Bear, on the house for all your efforts, Doc."

Malcolm squeezed Jolene. "You know Shannon, it's been such a long day, I think I'll take a rain check."

Shannon shrugged. "Suit yourself."

*****

Malcolm straggled into his house, released Precious and her rotund offspring out the back door and stepped out onto the patio. The unassuming twilight of the backyard was ideal to decompress from the chaos of the day. The puppy stumbled after a rabbit, which zipped under the fence, leaving the confounded stalker to sort out the vanishing act. Malcolm turned to Jolene. "He's the one that most closely resembles Precious. I'd like to find a special place for him, since Delmar and Henry have adopted their new arrivals."

Jolene responded, "I'm surprised she didn't pay any attention to the rabbit."

"Still recovering from the spay surgery last week, I imagine."

243

Malcolm looked about and found her by the opposite fence, nosing the ground. Malcolm appraised her more carefully. She's not sniffing, she's eating something. "Looks like she found some rabbit pellets."

He called out, "Precious, leave that alone!" Precious looked up, chewing on something substantial in her mouth. Malcolm strode across the yard. "Damn it, it's probably some sort of rotted carcass."

Jolene asked, "What would *that* be doing in your yard?"

Precious continued her consumption of the mystery treat with alacrity.

"Precious! I said leave it be!"

She backed off as Malcolm bent over and picked up a paper plate by the bottom rail of the fence. He examined the texture of the contents on the plate carefully. "It looks like fresh ground beef." Malcolm scanned the tree line behind his back fence. Uneasiness gnawed at the edges of his mind.

"What is it, Malcolm?"

Malcolm returned to the patio, shaking his head. "Something isn't right."

Precious was at his heels, hoping he would change his mind over the confiscated contraband. Entering the house, Malcolm turned on a light and sifted the ground round with his fingers. The beef had a gritty feel, like sand added to plaster. He pinched a piece of the mixture and crushed a small green pellet between his finger and thumb. His hair stood on end. "Damn it! This shit's been laced with bait, she's been poisoned!"

"Poison!" Jolene grabbed the flimsy plate. "What kind of poison?"

"Probably some rat or varmint bait." Malcolm clutched her upper arm. "Get the puppy in from the yard please, I don't trust him to be out there alone. I've got to find something to make Precious vomit."

Malcolm tore into the garage, fumbling and cursing through the disarray in his grip. He didn't bother shutting Lucille's back hatch or the kitchen door as he rushed back into the house, nearly knocking the puppy out of Jolene's arms. Precious sat under the table, still hoping the treat on the paper plate would be relinquished back to the floor. Malcolm bent over and popped the injection into Precious' leg.

"How long before that works?"

"A few minutes. I've got to take her to the clinic, I need fluids and other things I don't have here."

"I'll go with you."

Malcolm had a mind to protest but decided discretion was the better part of valor. "We better take our young friend as well."

Malcolm bore Lucille down the valley highway, paved in silver from a half moon directly overhead.

Jolene cradled the puppy. "Who do you think did this?"

"Some bastard who knew where I would be all day; it would be easy to slip the bait under the fence when no one was around."

"If we had gotten home a little later when there was no light outside and hadn't seen her eating from that plate…"

"I'd rather not think about that right now."

Precious' abdominal muscles contracted in waves and with a gagging heave, she spewed undigested meat, covered with a tacky gravy of green dye and bile, onto the back seat.

"That's a start," Malcolm commented.

"You think that's all of the poison?"

"I doubt it."

Malcolm wheeled Lucille up to the clinic and jammed the brakes to a jarring halt. Henry's Jeep was parked in front. Malcolm hopped out of his seat and opened the rear door. Precious was slumped against the back of the seat; her front legs rigid as if trying to push herself off the cushion. Malcolm scooped her up and directed, "Jolene, carry the puppy in!"

Banging into the office, Malcolm came across Henry and Lois, bent over a pile of catalogs on the desk. Lois exclaimed, "What's going on?"

Malcolm went straight to the back room, responding in a terse voice, "Precious has eaten poison bait. She's starting to show seizures of her extensors."

Jolene entered the scene with the puppy and shared a mutual long regard with Lois. "Hello, Lois."

Lois smiled, "Looks like we're both helping on the rounds this evening."

Malcolm rested Precious on the exam table and grabbed a set of clippers. Precious' legs tightened into a sawhorse posture, her body trembling slightly.

Henry said, "Get that vein clipped, Malcolm, I'll draw up the pentobarb. Lois, get a bag of fluids ready."

Malcolm inquired, "Pentobarb?"

"To counteract the muscle seizures, it looks like strychnine. Did you try to get her to vomit?"

Malcolm forced himself to control his breathing. "Yeah, but apparently it wasn't enough."

Malcolm held Precious while Henry slugged the drug into the vein. Henry stepped back, "Okay, we'll probably have to repeat that, let's see if we can get anything else out of her stomach and start some fluids to flush the toxin through her kidneys."

Precious legs softened. Malcolm gently stroked her head, whispering, "Hold on girl. You damn food hog, just hold on."

SPRING

CHAPTER XLIII

Wheat awakened, cast off winter's blanket, and stretched in emerald ribbons across the valley. Plows ruptured a new covenant of fertility between farmers and earth.

Tramping outside by a barn wall after extracting twin lambs from a ewe, Malcolm spotted crocus blooms brazenly defying the chilly air. "Spring is finally here."

"Not until after the *Zwiebel-Schnee*," Nahum Y. Peachey replied.

"*Zwiebel*, as in onion?"

"*Ja*, onion snow. After the onions shoot up in the garden, we get one more snow, the *Zwiebel-Schnee*. Then spring stays."

Malcolm contemplated the moods of the weather as he filled his bucket in the milk house with fresh water. Opening the door to wash his boots under the soft morning sun, he came face to face with a young woman, perhaps nineteen years old. The girl averted her eyes and blushed, Malcolm gaped and froze in place; both thrust into an uncomfortable space. Their last encounter had been just as awkward when he came across her and Samuel Y. Peachey in the straw of his father's barn.

A withered figure emerged from behind the girl, shattering the spell of immobility. Zephaniah Y. Peachey's eye drilled into Malcolm and subsequently targeted the girl.

Malcolm nodded and slipped away from the milk house portal. Nahum Y. and the girl stood still, their hands clasped in front and heads lowered. A deep cannonade roared from the stable, as the bull voiced the patriarch's silent admonition. Malcolm brushed his boots, dumped his bucket and escaped into Lucille. He exited the scene, preferring that the remaining characters would act out the dark pageant of repressive theocracy on their own.

Decompressing from his day, Malcolm leaned against the patio wall. Precious and her rapidly growing puppy wrestled in the fresh green grass of the backyard. *Wonder who would provide a good home for the last pup?* He hit upon an idea. The phone disrupted his meditations and he managed to answer by the third ring

"Hello, Malcolm," greeted Stephanie Teesdale.

Malcolm realized that sooner or later, he would have to accommodate her visit. He volunteered his newfound knowledge of local cave lore, acquired from the Dead Bear sages.

She exclaimed, "That's super, especially if you can find out the whereabouts of this Dark Spring Cave in relation to Mr. Yoder's farm. We could explore that cave and others in that area."

"When are you thinking of coming up here?"

"Don't know exactly, but probably within the next three to four weeks, the bats should be fully active by then. I'll have a colleague from the university with me, a bat biologist. He may want to collect blood from some bats for serology and check titers for rabies. Can we pick a date that you can accompany us?"

"I'll try."

"That would be great. Have you run across any more cases?"

"Of rabies? No."

"I'm looking forward to meeting you. Thanks again and have a good evening."

"And you as well." Malcolm hung up the phone receiver onto the wall. *How do you trap bats?*

He pondered who might best serve as a guide or know of one. Jolene and Bobby weren't at all familiar with that area of the valley, Chief Cal was hardly an option. He had no choice but to ask Shannon, a bad omen for the adventure if there ever was one.

*****

Malcolm felt as if ten years, rather than one, had passed since he first blundered into this pasture and all the hardship that came with it. Navigating the rabies cases and the peculiar attitudes of the valley had proved to be difficult. At times, he seemed to be wandering in a labyrinth, without direction or destination. Malcolm suspended his thinking. "I guess things have settled down for you this past while, Jonas."

"*Ja Tierarzt*, it is better." There was a pause. "Do you think that *mein lieber* Kip had the rabies, then?"

Malcolm's heart sank, weighed down by the bitter cost of Kip's loyalty for his master and his sheep. "Unfortunately Jonas, I believe that is a possibility. We'll never know."

Jonas gazed into the distance. "I miss him."

"He was a good dog." Needing to shed the depressing images assembling in his mind, Malcolm inquired, "Do you think I could get a two dozen eggs?"

Jonas Y. cracked a thin smile. "I'll get Rachel."

Malcolm loitered by Lucille as Rachel and Jonas P. came out of the milk house. Malcolm opened Lucille's back door and hoisted the last puppy from the seat. Rachel's face warmed, while Jonas P. rubbed his beard.

Malcolm explained, "He's kind of a big brute for his age, three months already. I thought you could use a new friend. He's got a great heart and he's part cattle dog."

Rachel laid the eggs on the hood of Lucille and stroked the cocoa colored head.

"He's had all his shots and has been wormed, Jonas. I think he would be a good pal for Fritz to keep an eye on the farm for you."

Rachel pleaded, "Daddy?"

The farmer nodded. Malcolm gently rested the puppy, wagging his tail and licking Rachel's face, into her arms. Her arms sagged but held her new friend firmly.

Malcolm retrieved his eggs. "Take good care of him."

Rachel watched the young veterinary drive off. She hugged the soft mass of fur and set him on the ground. She winced and grabbed her left shoulder; waves of numbness pulsed down her neck and back, followed by a touch of dizziness as she stood back up.

"*Was ist das Problem*, Rachel?"

"*Kein Problem*, Daddy." The puppy playfully chewed on her bare feet.

*****

Delmar sat in the living room, slumped in an armchair, reading the paper and stroking the canine lump on his lap. Jolene reposed against an arm of the couch, drinking coffee and taking a break from the spring cleaning of the heifer pens.

"Looks like Duke is breaking you in, Pop."

Delmar rolled the puppy's ears between thumb and fore finger. "Can't get him to retrieve anything worth a damn."

"He's part cattle dog, maybe he'll be useful in the barn."

Delmar peeked over the paper at his daughter. "Yeah, I hope a good deal more than that bonehead Pierre." Delmar returned to his research of the day's news. "See this Jolene? They've got real photos of UFO's out west." He lowered the paper for emphasis. "I wonder if them aliens would take a shining to the hills around here?"

Rubbing aching forearm muscles fatigued from lifting forkfuls of manure and straw, Jolene answered, "Don't know, Pop," thinking, aliens would be a good fit for around here.

"Humph." Delmar turned the page. "Hey, look at this. Fella by the name of Haumiller got in a bad wreck last night, couldn't slow down his truck on the bypass by the reservoir, critically injured, so he is." Delmar laid the paper on his lap, blanketing the resting puppy, and regarded his daughter from across the room. "That's a steep grade on that curve. Not a good place for your brakes to fail."

Jolene stopped her self-massage and jerked straight up in the couch. "Can I see that article, Pop?"

CHAPTER XLIV

"This piece of shit truck ain't going nowhere," mumbled Lester MacDougall. Frustrated, he clubbed the non-functional distributor cap with his socket wrench. The search for a new part wouldn't be easy, given the age of his father's Chevy truck. He pulled down on the creaky hood and pounded it shut. Wiping his hands with a rag, he sauntered over to the back porch. John MacDougall, hair falling in his eyes and arm twitching, struggled to stack peanut butter and bacon sandwiches on the picnic table.

"Pa, I gotta go find a part for your truck."

MacDougall scooped a glob of peanut butter from a jar.

Lester watched his father smear the brown goo on a piece of white bread. His sweatshirt and pants seemed to be two sizes too large. "Pa, I'll have to come back later to finish the truck. Okay?"

The bristly head nodded in reply.

"Why are you making so many sandwiches?"

MacDougall halted with another knife full of peanut butter in midair, hand shaking. He raised his head to reveal droopy lower lids, melting away from blood shot eyes. "Leave me be! They're coming after him and he needs grub."

"Coming after who, Pa?"

MacDougall canvassed the woods beyond the clearing. "He's nearby, don't you know. Needs a place to hide, everyone in town is after him."

Lester marked the number of sandwiches on the table, at least fifteen. "But Pa, Uncle Archie is ..."

"No one will take my kid brother from MacDougall land, not the law or them God damned dutchmen!"

Lester collected his tools by the truck and headed for the barn.

As he opened a small side door, his father yelled out, "Careful ya ain't followed! Ya hear me, boy?"

Lester's shoulders sagged. He's gone bat shit crazy. Did he need to find a shrink to examine the old man? How would he ever get him to an appointment?

✳✳✳✳✳

All across the valley, farmers joined the annual crop lottery, hoping that the growing season would bring ample rain and warm sun, no damaging storms and no untimely frosts.

John Y. Peachey and sons were adjusting harnesses on a quartet of sturdy Belgians, rekindling the leviathans with a sense of purpose after the long winter. Noting the horse's spirit, the farmer advised, "Eli, don't put Jake and Tim together, this early in the season they might get after one another. And keep Pete on the outside, we need him to lead on the turns."

The farmer looked out across the field where he intended to drill the oats, plotting his driving strategy with the team. The sun was stretching well beyond the morning horizon, and daylight during planting season was always too short. "Samuel, Eli and I can finish with the horses, start bringing out the sacks of oat seed so we can load the planter." As his eldest son withdrew, John Y. scoped a black top buggy turning into the far end of his farm lane.

A Standardbred pulled the buggy down the dusty track at a fast clip. Shielding his eyes with his hand, John Y. recognized the driver. His cousin Nahum Y. Peachey held the reins, and unless he was mistaken, Nahum's father was with him. John Y. frowned. This is a bad sign for the spring planting.

Nahum Y. directed the horse into the barn yard and pulled up a few feet from the horse team.

John Y. waved, "*Gut morgen*, Nahum *und* Zephaniah."

Nahum Y. remained silent, staring at the back end of his horse. Zephaniah Y. fumed, his lower lip quivering.

John Y. asked, "*Was ist los?*"

253

The elder rotated to face the inside of the buggy. *"Gehen!* Whore! Go!"

A young woman slunk out of the buggy and gingerly set her bare feet on the ground. Welts blemished her ashen face, hooded under tangled hair that lacked the cover of a bonnet.

Samuel Y. returned with his burden of seed. The youth stopped, the sack falling to the ground and splitting open. "Anna," he murmured. Tears welled up in her eyes.

Zephaniah Y. sprang up, nearly pitching off the buggy. *"So, es ist du! Du bist ein schmutziger Hund!"* A flash of red colored the diviner's brow, his opaque blind eye rolling in its socket. He pointed to John Y. "If a man finds a maiden that is a virgin, which is not betrothed, and lay hold on her, and lie with her, and they be found; then the man that lay with her shall give unto the maiden's father fifty shekels of silver, and she shall be his wife; because he hath humbled her."

Zephaniah Y. fell back into his seat, barking, "We go now!" His son pulled the reins and the buggy spun into a tight turn. With a snap of leather, the horse broke into a brisk trot. The elder brandished a whip and set upon the horse, causing it to bolt at a canter and whinny.

John Y. watched his uncle's departure, took off his hat and scratched behind his ear. Resetting his hat, he surveyed the girl, then the sack at his son's feet. "So Samuel, it seems we have to clean up after the spilled seed."

*****

Given the poor suspension of his police cruiser, Chief Cal's ride on the rutted lane that led to the MacDougall place was particularly annoying. Yet, as he drove into the clearing, the mild weather managed to cast a glossy lacquer over the dour lodge. Cal made for a white Chevy pick-up with the hood up. A figure bending over the engine lifted his head long enough to frown at the sight of the police car. Cal rolled to a stop and hoisted himself out onto the ground.

254

His leg was bothering him again; if he was in Florida, he could forget about begging for a new police car from the township.

"Hello," Cal called amicably.

The back-yard gear-head leaned away from the hood, topaz eyes reflected coolness. His brown, curly hair was cut short on the top and sides but touched the shoulders in back. "Morning."

Cal took a second look, for an instant he faced a younger, longer haired version of John MacDougall. "It is a nice morning at that. It's been awhile, but I gather you must be Lester MacDougall, John's boy." No reply was forthcoming. "I recollect you left your daddy's home a while back to go work in Gillstown?"

"That's right."

"You're a mechanic?"

"Yup."

"You may not remember me, but I'm Chief Cal Geisfelt. Thought I'd stop by and see how things are going."

Lester remained in place. Cal sauntered over to the truck. "Keepin' the old girl up and running I see."

His host gestured with a wrench in a grimy hand. "Ain't been starting right. Thought it was the distributor cap. But now I think the fuel pump needs fixed as well." As he spoke, a gold tooth glimmered in the front of his mouth.

"Lots of room to work under those old hoods. Is it a '64?"

"No, '63."

Cal chuckled, "Things were simpler back then, no convoluted electronics. Still got the original engine?"

"No, rebuilt 350."

"That's a little kick under the hood." Cal ran his hand across the back of his neck. "Is your father around, Lester? I would like to see him for a minute or two."

The wrench revolved in Lester's hand. "Been off in the woods."

"Oh? Finding a spot for the upcoming gobbler season is he?"

His eyes drifting away, Lester swallowed. "I imagine so."

The chief set his hand on the truck's cab. Gazing steadily at Lester, he asked, "You living here now?"

"Nope, just helping Pa with his truck. I'm handy , I guess."

"Good skill to have, son." Pushing away from the truck, Cal regarded the sunlit yard behind the lodge. "Anyone else from the family visiting these days?"

"No, sir."

"Fom the looks of things on the table, you folks are fixin' to have a picnic."

"Pa likes to feed some of the wildlife."

"Like peanut butter sandwiches do they?"

Lester nodded. "Yeah, deer and squirrels."

"Is that a fact? Didn't know squirrels were partial to peanut butter. And is that bacon I see on that plate? Hope you don't attract bear." Cal scanned the pasture behind the barn. His eyes halted on the tree line to the west of the barn. "So your father isn't around right now, then?"

"Afraid not."

Cal rubbed his chin, letting the moment pass unhurriedly. "Hard to believe you haven't had a bear visit your buffet. They sure have an appetite for such things."

Lester jerked a thumb towards the truck. "I guess I should get back to my work."

Cal held up his hands. "Of course, of course. I appreciate your time." He took one last look about the clearing and retired to his Ford. Two crows burst out of the woods and rocketed into the air, calling one another in short squawks. Refraining from getting into the car, Cal beheld the black duo as they skimmed the treetops and were joined by four others as they ascended the ridge spur that reached down into the clearing. Lester clenched his wrench with enough force to transform his weathered knuckles into plaster.

A disheveled figure lurked by the corner of a kitchen window, watching Cal loitering by the truck. Smacking his tongue and lips on peanut butter that he fingered into his mouth, he wondered what Lester was telling the law. Jaded eyes studied Lester—betrayal always came from where you least expected it, even family.

CHAPTER XLV

Precious wedged herself between the door and Malcolm's legs. "I know it's a weekend but you're staying here!" He squirmed past her, shut the door and marched towards the Scout to find Jolene, arms folded, leaning against the hood.

Precious appeared in the kitchen window, protesting strenuously. Jolene said, "Looks like she's got herself up on the table."

"I don't know what her issue is this morning. She usually is fine when I have to leave without her." Malcolm yanked the driver's door open and turned the key. The engine coughed and groaned. Malcolm rapped his fist on the steering wheel.

Jolene stuck her head through the passenger window. "International Harvester, reliable as ever."

Malcolm compressed his lips, pumping the pedal and trying the ignition simultaneously.

"That's why you don't see a red tractor on my dad's farm."

"Yeah, but this is a Scout, not a tractor." Malcolm wilted into the seat. "Nice timing, Lucille."

Jolene inclined her head towards her truck. "Come on, we'll take my ride."

Malcolm lurched out of his seat and slammed the door. "We could have walked and been there by now."

Jolene wheeled in front of the Dead Bear to find Shannon, Jimmie, and Ronnie dawdling in the parking lot, smoking cigarettes and drinking beer.

Ronnie held out a pair of green bottles. "Would youens like a

Rolling Rock?"

Malcolm held up a hand. "No, thanks. I'll need all my energy to tromp around the woods and caves today."

Ronnie shrugged and imbibed from one of the bottles.

Shannon asked, "So youens want to have a gander inside Dark Spring Cave?"

Malcolm replied, "I think that's the plan. And depending on time, perhaps other caves near it, the crew from the university is coming up here and I figured it might be good for me to learn a little about rabies from them."

"You know your way up there, Doc?"

"Not really Shannon, but Dale and Jack Calhoun said they would guide us."

"Good thing that, not the kind of place you want to get lost, that's why I asked them on your behalf."

Ronnie piped in, "You never know what's up in them hills."

Malcolm asked, "Why don't you guys come into the cave with us?" Jolene raised an eyebrow at his suggestion.

As if choreographed, the trio all took a step back in unison. "Not on my life," Ronnie said. "I don't go into caves, and I don't spend time up in that part of the valley."

Shannon inquired, "I don't see your dog in the car today, Doc. Word got around she got into some poison bait."

Malcolm deduced the origin of the news leak—Lois. "I think she's made a full recovery, Shannon, thank-you."

"Hell of a thing. There you were, helping out all those folks at the fire hall and someone goes and dumps shit like that on you."

Jolene's eyes blazed. "That wasn't some shit, that was pure evil."

Shannon jetted smoke from his nostrils and creased his brow. Smiling, he replied, "You have a point there. Although evil doings often lead to inconvenient consequences for the evil doers."

Jolene tried to read Shannon's poker face. Her concentration was broken by a blue 4x4 Ford scudding to a stop; barely three feet away from Malcolm. Perched in the passenger seat, Dale Calhoun took stock of the gathering. "Guess we're heading up in the hills for a hunting party today. Maybe we'll run across that potlicker brother of MacDougall's."

"Or perhaps his black dog," Malcolm said. Jolene brushed Malcolm with a sweep of her eyes.

Dale raised his gaze as if deep in thought. "Yeah, that could be a possibility, too." Jack tapped his fingers on the steering wheel.

Malcolm continued, "I guess I shouldn't have worried about that beast having rabies after all, on account of him not causing any trouble lately."

Dale nodded. "Yupper, not much."

Jolene asked, "I thought you guys might have asked Chief Cal to supervise the cave search."

Dale turned to his brother. "Didn't see his car at the township hall this morning. Did you, Jack?"

"Nope."

Dale said, "Don't need him around to interfere, anyhow."

A newly-minted red Jeep Grand Cherokee rolled into the town square, balked, then edged towards the band of adventurers. As the vehicle rolled to a halt, long legs in cargo pants, unblemished hiking boots, and an L.L. Bean jacket glided out of the driver's door. A bearded middle-aged man, with glasses and baseball cap, remained in the passenger seat. Behind him sat a younger woman, absorbed in a fistful of paperwork.

The new arrival asked, "Hello, I'm looking for Dr. Malcolm Cromarty." All other conversation ceased.

Malcolm raised his hand. "That would be me."

The safari blond reached out her hand. "I'm Stephanie, Malcolm. I'm very glad to meet you." Her fingers laced his palm in a relaxed handshake.

"Nice to meet you." Malcolm introduced her to the Calhoun brothers and Jolene.

Stephanie considered Jolene. "You're a veterinarian, too?"

"No, local journalist."

"Reporting on science for the community paper? How nice."

Muscles tightened across Jolene's face. Leaning on the

passenger door, Stephanie lowered her head to better view Jack in his truck. "Why, this truck looks like it could go anywhere in these hills, I'm sure you're the type of guy who…" She stopped in mid-sentence as Shannon unabashedly reached out a hand and introduced himself.

Stephanie recovered her poise. "It's very kind of you to want to come along, but this will be a simple affair."

Shannon turned up the charm to his highest setting. "Doctor Stephanie, much as we would like to accompany you, we'll hold the fort back here. But you're more than welcome to stop by for a cold one on the way out of town when youens are done with your bat collecting. You can have yourself a great burger, too, with all the fixins—on the house."

Stephanie forced a smile. "I'll consider that when we're finished, don't know how long we'll be though, or even how far we're driving."

Malcolm turned to Dale. "Do you know the best way to get to the cave?"

Jack brought his truck's engine to life. "Just follow us."

Stephanie stretched an arm to the Jeep. "Perhaps you'll ride with us, Dr. Cromarty?"

"Can Jolene ride along as well? There's not much gas in her truck, and my vehicle wouldn't start this morning."

Stephanie wavered and eyed Jolene. "I'm sure we can accommodate you."

"Thanks so much," Jolene reciprocated, with a smile that didn't include her eyes.

Stephanie spun about and quickly paced to her Jeep.

"Bon voyage," Shannon called out, raising a beer in hand.

Jolene took a quick glance over her shoulder. "Malcolm did you see that article in the paper about Haumiller having that serious accident in his truck? It seems his brakes failed on the hill up by the reservoir. He's in critical condition."

Walking next to her side, Malcolm said, "No, I didn't see that. Is this the same Haumiller that I know and love so well?"

"The same one. Kind of a strange occurrence, don't you think?"

"I guess it's possible brakes could fail on that hill. It is pretty steep. I'm not much of mechanic, though."

"I'm thinking more saboteur than mechanic."

The dual carburetor on Jack Calhoun's truck pushed his followers to move along.

Malcolm and Jolene approached the Jeep. Stephanie addressed the man in the baseball cap. "Dr. Kincaid Sinclair, this is Dr. Malcolm Cromarty, but call him Malcolm, and his friend, uh..."

"Jolene," Malcolm inserted hastily.

"My pleasure," responded Sinclair as Malcolm and Jolene entered through the rear door. Malcolm squeezed into the middle of the back seat. The young woman looked up from a map. "I'm Lydia, Dr. Sinclair's grad student."

"Hi, I'm Malcolm, and this is Jolene."

"So, you're the vet?"

"Yes. What do you and Dr. Sinclair study?"

"Chiroptera."

"Pardon?"

"Commonly called bats."

"Of course." The doors closed and they lurched after Jack's truck.

"That's an interesting group of, uh, acquaintances you collected for our arrival today," Stephanie remarked.

Malcolm shrugged. "No collecting necessary, they appear on their own."

"Did that short guy just leave a bowling game?"

"Perpetually, it seems."

The Calhoun boys roared up the Shaneytown Road towards the misty hills that edged the eastern end of the valley. A gray veil masked the sun and a breeze stirred the humid air. As the motorcade split off on the familiar right leg of the "Y", Malcolm said, "We're coming up on the Yoder farm where the sheep were attacked, and beyond, the MacDougall place."

Stephanie queried, "What happened to the dog that attacked the sheep?"

"I understand it died."

"How?"

"Accident…perhaps." Sinclair and Stephanie appraised one another.

Lydia unfolded another map. "Could you identify these two places on here please?"

"This seems to be fairly detailed." Malcolm perused the paper. Getting his bearings, he fingered the spots for Jonas P. and MacDougall. Lydia wrote numbers in red ink and circled them. Malcolm tapped an additional numbered circle. "Another place of interest?"

"That's where another rabid dog infected with the bat variant of the virus was discovered by animal control officers."

Malcolm peered at the location on the map. "A border collie perhaps?"

"Don't have that info here."

"So that dog had the same strain of rabies as the sheep?"

Sinclair cleared his throat. "Indeed likely the same variant that infected Mr. Yoder's sheep."

Malcolm's heart sank.

Sinclair continued, "We are starting to realize that different variants are not only present within mammalian orders, such as Chiroptera and Carnivora, but also within families. That is, species-specific variants. We haven't tested it yet, but we surmise the dog might have had the Ln/Ps variant, the one that is most commonly associated with human cases."

"Ln/Ps variant?" Malcolm asked.

"*Lasionycteris noctivagans* and *Pipistrellus subflavus*."

Stephanie said, "The Silver-haired and Eastern pipistrelle bats." Malcolm hardly knew the species, even in English.

Sinclair added, "Probably the most common occurrence of intra-species rabies infection occurs among *Eptesicus fuscus*, but we don't believe this variant has been found to be causal in humans very often."

"Big brown bat," Lydia whispered.

Malcolm thought. At last, a bat I recognize. "Why do big brown bats have more problems within their own species?"

Sinclair partially turned in his seat. "Well, *Eptesicus* specimens are submitted most frequently, hence the higher rate of discovery of infected animals. Also, they are a sedentary, colonial species. Their habitat includes chimneys, attics, or other man-made structures like old, abandoned buildings. They remain all year and hibernate in cold weather as one large group. You have a clustered population, territorial disputes arise, aggressive behavior follows and given that the virus can incubate for months before clinical signs, it can maintain a high prevalence in the population."

"Did you say months?"

"Yes, of course."

Jolene kept her focus on the scenery out the window.

Malcolm probed, "And the other bat species?"

Sinclair chuckled, "Well, our diminutive friends migrate a bit more, although we aren't sure if *Pipistrellus* travels great distances. They are colonial, but *Lasionycteris* is a more solitary fellow. Both of them are likely to be found in places like old trees, hollows, perhaps small caves or grottos. If Mr. MacDougall's dog had rabies, and that is a big *if*, then perhaps it encountered some of our winged friends in a cave."

The canopy of leaves over the road further dulled the murk under the gathering clouds. Stephanie veered the Jeep into a side road that ascended the ridge. She found time to glance at Malcolm in the mirror despite the curvy track. Jolene observed the reflection and sighted Malcolm out of the corners of her eyes.

"You see, Malcolm," Stephanie said, "What makes this unusual is the "spill-over" effect. That is, the bat variant of rabies was found in the ram, with the possibility that something other than a bat, like the black dog, acted as the transmitter. This secondary transmission is not thought to occur with the bat variant very often."

"So, Jonas P.'s ram was infected for a long duration, like infected bats?"

"Yes, at least until it showed clinical signs. Bat variants of rabies in other species usually causes fatal encephalitis so quickly and clinical signs develop so fast that no transmission can occur. So non-Chiropterans are what we call dead-end hosts."

Malcolm digested the lesson…or just plain dead hosts. "So, what are we trying to accomplish this afternoon?"

"It would be useful to find any evidence of bats in the area of this cave, given the proximity to the MacDougall homestead."

Sinclair jumped in, "We probably won't readily discover any *Lasionycteris*, the little imps. But if we are fortunate, perhaps we'll find a small colony of *Pipistrellus*. We can capture a few and collect blood for rabies serology, perhaps return to the cave if we identify positive samples and collect some specimens for virus isolation and genetic typing."

"Specimens?"

"Why yes, some of the bats for necropsy. Think of the great opportunity to find a previously unknown variant of rabies virus."

Stephanie kept pace with Jack's truck into a lane that was little more than two dirt tracks liberally dotted with fist-sized rocks. Obstinate branches scratched at the windows, squeaking in protest of being pushed aside by the Jeep. They came upon a glade to discover the Calhoun boys already out of their truck. Jack was lifting a shotgun from the rack behind the seat. Stephanie eased beside the pickup.

Sinclair pushed open his door. "Now the search begins." He no sooner dangled a leg onto the ground before rebounding back into his seat. "Good God! Is the gun that the guide is carrying necessary?"

Malcolm pondered over an answer. "Have to be careful of bear this time of year. Moms out with the new cubs."

Jolene's lips twitched with a subtle smirk.

Dale tapped on the driver's window to gain Stephanie's attention. He pointed to a path that trailed along a small run that bubbled over rock and moss on its way down the slope. "We have to hoof it about a hundred twenty yards from here."

Stephanie stepped out of the Jeep. "Lydia, will you please get the packs?"

Hands on hips, Dale watched Lydia gather equipment for the march to the cave. "Going to catch some little bats with that netting?"

Sinclair retorted, "If we find some 'little bats', as you call them, we shall endeavor to capture specimens for a diagnostic workup."

"Good thing we've got scientific minds on the job."

Malcolm contemplated Dale's expression, it suggested sarcasm.

Grabbing one of the packs from Lydia, Stephanie called out, "We're ready."

The smell of damp bark and soil permeated the narrow path. There was little sound except the occasional bird or chipmunk chattering among the wild blueberries. The climb wasn't steep, but the humidity pulled beads of sweat from Malcolm's temples. The passage leveled onto a flat, with a commanding vista of the valley opening to the left. To their right, a dark void over thirty feet wide and twelve feet high, gaped in a wall of lichen-dappled rock.

Dale pointed to the south. "Yeah, less than a stinkin' mile and a half from his place if you cut back around that spur of the ridge and into the next hollow."

Lydia relieved herself of her pack and opened her map. "Which place is that?"

Dale huffed, "Them that started all of this mess."

Malcolm said, "He's talking of John MacDougall's place, but that would be a tough hike over the ridge from here."

Thunder vibrated off the ridge tops and agitated a light drizzle out of the clouds. Primal instinct prompted the party into the shelter of the cave, with Sinclair leading the way.

Jolene held back with Malcolm to tag along in the rear. Squeezing her forearm, Malcolm said in an undertone, "This is getting a little creepy."

"Don't worry. We have experts to guide us."

CHAPTER XLVI

Shards of maple spun through the air with each blow of the axe. Jonas P. paused to wipe his forehead. His sweat was worth the effort. His wife needed the wood stove heated up to simmer a new pot of ham and bean soup, one of his favorites.

Fritz bolted past him towards the sheep paddock, leaving the awkward puppy to follow in his wake. Growling and hair on end, Fritz oscillated along the fence of the enclosure. Jonas P. dropped his axe and jogged into the house. By the time the farmer reappeared outside, rifle in hand, Fritz was beside himself as the flock of sheep circled from one end of the pen to the other.

Crouched within the paddock, a gaunt, naked figure tracked Jonas P. with sallow eyes. The intruder raised a stained hand overhead, brandishing a sickle.

Jonas P. yelled, "*Was ist los? Was soll das?*" Mindful of his dogs and sheep, he fired a round into the air. The phantom loped behind the shed as the second shot tore a chunk off a fence post.

Jonas P. neared the fence, his pulse racing up the sides of his neck. Despite the frenzied barking of the dogs, wheezing gasps could be heard behind the shed. He sidestepped to the left with uncertain steps. A shadow materialized from behind the corner of the shed and flung a fistful of mud and gravel at the farmer. Jonas P.'s brimmed hat fell to the ground as he jerked his head and covered his face from the debris. Regaining his grip on the gun stock, Jonas P. went to take aim but had no target. Time suspended as he poised to rebuff the next assault, it didn't come.

Finding his nerve, Jonas P. skirted about the corner of the shed.

Fleeing across a small clearing, the berserker had nearly reached the hardwood cover at the foot of the mountain. The farmer shouted, "*Gehen* you devil!" and squeezed two shots. His target contorted into an uneven gait but managed to reach the tree line. Clutching its left arm, the marauder looked back at the farm, reeled a screeching howl and vanished into the brush.

Hands shaking, Jonas P. could hardly constrain himself as his wife and children poured out of the house. He shouted, "*Zurück im Haus!*" Rachel stood her ground as her mother hustled her siblings through the porch door. She stared at the deep forest stretching into the ridges, listening for the howl to return.

*****

Lydia conjured up a couple of fluorescent lanterns and handed them to Sinclair and Stephanie. The entrance of the cave widened into a chamber twenty feet high and nearly a hundred feet long. The ceiling sloped downwards to the rear, requiring any curious seekers to crawl on hands and knees to touch the back wall. Two fissures, their jagged borders wide enough for human entry, stretched from the ceiling to a shelf of rock six feet from the floor along the right wall. Water seeped from one of the cuts, tracing a charcoal gray swath across the shelf, onto the floor, and out the cave entrance.

Raising his lamp to gain perspective, Sinclair mused, "I wonder where those smaller passages go? Perhaps we should climb up on that ledge and get a better look."

Dale scrutinized the clefts. "You mean by creeping around in them?"

"Yes, a little spelunking might be in order."

Malcolm silently agreed with Dale's sentiment. *Better him than me.*

Stepping on a foothold in the wall, Stephanie stretched to gain a better view of the ledge below the crevices. "Dr. Sinclair," she pointed to a mottled pile of debris.

Sinclair took a couple of steps. "Ah, it seems our quarry has been here."

"Bats?" Dale asked, searching the rock ceiling. Sinclair was too lost in thought to hear.

"Guano," Stephanie whispered.

Malcolm said, "Can't aerosol transmission of rabies occur in caves from exposure to bats?"

Dale pulled his collar tighter. "Well, there ain't much to impress me here, I think I'll step outside."

Leaning against a wall and cradling his shotgun, Jack agreed. "Let's get some fresh air. The rain has let off a bit."

The Calhoun brothers shuffled out of the cavern.

Stephanie shook her head as she watched them exit. "There's only tenuous evidence at best and *if* it happened, the circumstances were very unusual. Most of the human cases caused by bats are cryptogenic." Malcolm furrowed his brow. She added, "Cryptogenic, the affected person doesn't report any contact with a bat."

"How can that happen?"

"If a dog bites a person there is enough trauma to warrant medical attention, or at the very least, cleaning and awareness of the wound. But the *entire* mandible of some of the smaller bats like pipistrelles can be less than one centimeter in length. So imagine how small the teeth are. People waking up with a bat in their bedroom may not realize they were bitten, believing a day or two later the wound was a bug bite or thorn scratch."

"Sounds like vampires."

With a boost from Lydia, Sinclair huffed and puffed his way onto the mantle. Stephanie continued, "There's no need for hysteria about bats Malcolm, as long as there wasn't any contact. They're beneficial creatures."

Sinclair secured his feet and squeezed through one of the crannies. Lydia adroitly scrambled onto the shelf and examined the entrance to the matching gap. Jolene clicked on her own flashlight and crawled on all fours to examine the back wall.

Malcolm asked, "So just what sort of an extreme situation can lead to aerosol transmission?"

Stephanie's face lit up. "The kind of place that's right out of a horror movie." She observed Lydia vanish into the second chasm and returned her gaze on Malcolm. "There were a couple of presumptive cases reported back in the '50s. One was a researcher who was studying and handling bats, including banding and bleeding them. The second was a mining engineer looking for opportunities to harvest guano."

"Harvest bat guano?"

"High in nitrates and phosphates, good for fertilizer."

"Easier to get cows to fill manure wagons."

Stephanie shrugged. "Both of these men frequently visited caves, and one cave, Frio Cave in Texas, was especially notorious." Sinclair and Lydia remained nowhere to be seen.

Stephanie leaned closer to Malcolm. "Frio Cave was described in an article by a public health investigator by the name of Constantine. He reported that there were as many as twenty to thirty million bats occupying the cave, depending on the season."

Malcolm refused to believe his ears. "Did you say million?"

"Picture it, a seething, swirling tornado of wings. You could hardly stand without having bats collide into you. Bats massed so densely on the ceiling that urine and guano fell like rain below, forming a layer on the floor several feet thick. Imagine the ammonia and the stench arising from this sticky drizzle."

"Lovely."

"But it gets better. That's only the *bat* contribution. Where there's a high population density of mammals and waste, the arthropods are sure to follow. So there you are, a foul mist searing your lungs, having to stay fully clothed because swarms of flies would bite your face and hands. The mites were so thick that they formed plaques on the walls and ceilings, and would crowd around your eyes and mouth, nostrils and ears, causing dermatitis."

Malcolm scratched his ears and neck from phantom bites.

"But the most macabre aspect was the hordes of beetles that lived among the guano. They would prey on any bats that bumbled into this filth on the bottom, devouring them into nothing but bones in minutes, like terrestrial piranha. I guess they would attack humans as well."

Malcolm looked about his present environment. "Finally, the long-lost level of Dante's Inferno reserved for politicians."

Stephanie looked a little puzzled. "There was some debate as to whether the victims contracted rabies from a bite, especially the one that was handling the bats for bleeding. But even if there was aerosol transmission involved, you can understand what I meant by extreme conditions."

Lydia appeared from the grotto, brick-colored earth smeared on her face and hands. She slid down the face of the wall. "Not much there, just a small air shaft that probably reaches the surface."

Stephanie held her lamp above her head and looked up at the other crevice. "I wonder where your advisor is, Lydia?" She lowered the lamp, handed it to Lydia and strolled towards the cave entrance. "Well in the meantime, please gather up the gear into the bags."

Jolene stood by Malcolm's side, the knees of her jeans also soiled with a murky red stain. She said in a low voice, "Malcolm, there's something you should know." As Malcolm drew his attention to her, screams and shouts erupted outside the cave. Malcolm sped out of the cave, with Stephanie and Lydia in hot pursuit.

CHAPTER XLVII

Malcolm burst into a grainy drizzle to discover Dale and Jack twenty yards to the left of the cave entrance. Jack had chambered a round in his shotgun. A moan rose from a rock outcrop above their heads. "Help me!"

"Kincaid!" Stephanie dropped her pack and called out, "Kincaid, where are you?" Lydia was close behind.

Malcolm and Dale scrambled up a series of flat boulders to find Sinclair propped on his elbows, waist deep in a small sinkhole. "I slipped and fell back into this shaft as I tried to climb out and badly twisted my foot."

Puffing out of breath, Stephanie knelt beside him. "Are you okay, Kincaid?"

Dale and Malcolm each grabbed Sinclair under an arm and lifted him onto the surface, setting him down with his back against a tree. Sinclair winced as Dale removed his boot and examined his foot. Dale shook his head. "Pretty swollen, might be he broke his ankle."

Stephanie declared, "We've got to get him to a doctor!"

Gently manipulating the foot, Dale agreed. "It needs looked at. We'll have to take him for X-rays." Sinclair exhaled in rhythm as if preparing for impending labor.

"We need to go, now!" Stephanie demanded.

Dale looked up. "Relax, Miss, me and my brother will take him down to the firehouse and get the paramedics to give him a ride to the ER." Dale shouted to his brother who had remained below the outcrop. "Jack! We need to get the bat doctor back to town."

Sinclair grabbed Stephanie's hand. "Did you see them?"

"See what, Kincaid?"

"*Pipistrellus!* When I squeezed through the passage from the cave I found a shaft to the surface and stirred several of them into flight. I'm disappointed we didn't think to look here first, before we entered the cave, to set up our nets."

Jolene's face appeared over Malcolm's shoulder. "I guess we've done all the science investigating for today, and this rain doesn't make it any more inviting."

Stephanie glared at Jolene and scoured the ground to locate the jettisoned packs. Jolene presented them to her with outstretched arms. "Here you go, Dr. Teesdale."

Stephanie snatched the parcels and thrust them upon Lydia. "Could we *please* get going for Kincaid's sake?"

Dale and Jack helped Sinclair hobble towards the path that led to the vehicles.

Jolene drifted beside Malcolm and tugged him aside by his elbow. "I was about to tell you before everyone blew out of the cave. There's another passage in the back wall." Her hand unfolded. "Found these near the entrance of the passage, plucked them from what looked like a small fire pit."

Malcolm's fingers prodded the blackened bones of a small animal. He leaned slightly so as to be able to look past her to the cave entrance. She remarked, "I don't think this is a coincidence given our present location."

CHAPTER XLVIII

Milling about the open doors of the vehicles, the scouting party orchestrated the new seating arrangements for the trip back to town. Sinclair slouched across the back seat of the Jeep, his injured leg elevated on the backpacks that Lydia had obediently carried. She sat in the passenger seat, wet and tired, her eyes boring holes straight through the windshield. Stephanie said, "I don't see how we can possibly cram anymore passengers in my car given the need for Kincaid to rest comfortably. It'll be a bumpy ride down this track to the road."

Dale scratched his chin, trying to come up with a solution.

Jolene offered, "Malcolm and I can hang out here, you can come get us after you drop off Dr. Sinclair with the ambulance drivers at the fire hall."

Dale fidgeted, his focus oscillating between Jolene and Malcolm. "Don't much like this. You two shouldn't stay here alone."

Jack chipped in. "I'll stay here with them."

Jolene regarded the shotgun in his arms. "How is Dale going to get my truck back up here by himself? The both of you need to go to town so you can each drive a vehicle back here, we can't all fit in your truck with the four of us" She paused. "And I'm not riding in the back bed in the rain."

Dale said, "It ain't right."

Stephanie clapped her hands. "Let's go everyone! We have to get Kincaid to the hospital!"

Dale rotated as if standing on a slow-moving turntable.

"Alright, we're coming."

Jolene tilted her head in the direction of the Jeep and said in a low voice. "Better get going, someone is getting a little antsy."

Dale grimaced. "Damnation, we'll be back in a short lick. You keep out of sight until we return."

Jolene said, "The cave will be nice and dry." Malcolm pressed his lips together, deciding not to express his true feelings.

Doors slammed shut, engines came to life, and the small convoy maneuvered around the brush and took aim down the hill. The noise faded and solitude closed about them, not even a bird wanted to break the stillness.

Jolene said, "I imagine Dr. Teesdale won't enjoy her colleague's company in that new car of hers as much as she did yours."

"Who said I enjoyed hers?"

Jolene looked squarely into Malcolm's eyes and let him speculate on her silence. "Are you ready for this?"

"Ready for what?"

"To go check out the back passage of the cave."

"Go into a tunnel? With no way of knowing where it goes, or if it's a dead end?"

"I won't go anywhere that I can't get though easily," Jolene asserted. "Hell, when I was a kid I explored lots of caves and caverns with my brothers."

"To be honest, the idea of crawling under mountains and this one in particular, gives me the creeps."

Jolene pulled two lamps from underneath her jacket. "I don't think Dr. Stephanie will notice I borrowed these from the sacks. Now we'll have light."

Malcolm took one of the lamps from her hand and clicked the switch to make sure it worked. "You're full of surprises."

They retraced their steps to the cave entrance. She turned on her light and walked, then crouched her way to the cavity in the back wall. She thrust the torch into the space and sized up the diameter. Malcolm set himself on hands and knees and padded to her side. Holding up her hand, she instructed, "You wait here. If I have to slide back out, I don't need you jamming me."

"Be careful."

She touched his face. "I like adrenalin rushes."

"Not sure I'm sharing that sentiment right now."

Holding the light in one hand, Jolene squeezed into the passage. The soles of her boots dimmed with each slide of her knees. Soon after, not even her scuffling could be heard.

Malcolm sat with his back against the wall, content to have the grey light from the cave entry as his only source of illumination. The barely audible splatter of rain on leaves marked the passing of time. How far would she go before she thought she was stuck? What if she needed his help? His uneasiness ramped up with each passing second. A faint scraping, followed by an incrementally brighter glimmer, lit up the hole. Hands and arms came into view among a bundle of uncontrolled auburn hair.

Jolene was out of breath, but whispered excitedly, "Malcolm, this leads into another large chamber and perhaps a tunnel right through this ridge."

"What took you so long?"

She wiped her brow with her sleeve, put her hand on his arm. "I was only gone about two minutes." He wasn't convinced. "You only have to crawl on your knees for about ten feet; then there's a spot where it narrows so you have to crawl on your belly, maybe about a body length, take a slight turn left, get on your knees again, and then get up and walk!"

"Crawl on your belly? Sounds like a place to get stranded."

"You can do it, Malcolm."

"Just how tight is the narrow part of this hole?"

"If you're on your elbows, you might hit the top of your head. The entire passage is no more than twenty feet. Come on!"

Malcolm closed his eyes. "Alright."

"You wait here a minute, to give me time to clear. I'll see you on the other side." Once again, the lamp dimmed as her body filled the shaft. Malcolm consciously tried to concentrate on counting out a minute and turned on his lantern. He advanced one hand and knee, then the others in an uneven shuffle. His kneecaps complained as they slid along the unforgiving floor. The damp stone smelled of rusty iron.

The gap narrowed. Resting on his palms, Malcolm came to grips with his situation. Dear God, don't let this lamp fail. With a deep breath, he fully laid out onto his torso.

Pushing the light in front, Malcolm extended his arms and raised his hips to slide forward, curling his toes for traction. Dense air spilled into his lungs, inducing hyperventilation. What if Jolene had displaced all the fresh air in front of him and his body blocked what was coming from behind? Trapped, there was no other choice but to go forward. Each new extension and flexion of his limbs seemed to gain shorter distance. He rested; face down, hair matted from sweat and dampness. Asphyxia gripped his larynx. Amid the clamor of anxiety, 'Maybellene', the old Chuck Berry song, lifted from the vaults of his mind.

Encouraged by the audiopsychosis, Malcolm strained through another wiggle and looked ahead to discover that his light reflected off rock rather than empty void, the left turn. With the full extent of his fingers, he pushed the lamp around the corner; a forty-five degree obstacle to liberation. He rolled ever so slightly to his right side and snaked a hand onto the corner of rock.

He coiled around the corner. There, the ceiling descended to a little less than two feet above the floor. Laying his head on his extended arm, he could no longer see what lay in front of him. He slithered forward, his rear end brushing against the stone surface on top as he toed the floor. He gained another length of forearm and taking a breath of air, found it to be less stifling.

Scooting a little more, he rolled his head and glanced up, to see that the ceiling had risen slightly. Regaining his elbows, Malcolm exalted in his possible survival. After two more shimmies, he advanced into a crawl and was relieved to have his knees pound on rock once again. The passage took a slight bend to the left and there he beheld an aura of illumination about Jolene.

"You okay, Malcolm?"

He wobbled onto stiff legs and swayed a few steps, gasping large volumes of air into his lungs. He exhaled, "Coming out of that rat hole, you had a likeness to a vision of the Virgin Mary. I can't believe you did this without knowing what was on the other side."

I have a sixth sense for fresh air from all the hay tunnels."

"I'll have more empathy for the next calf I pull from a birth canal of a cow. I damn near suffocated in there."

"Look at the size of this passage!"

"This smacks of some sort of Hardy Boys adventure." He clicked off his light, deciding to save the batteries, just in case. "I could never be a coal miner. I need to regain my breath." Leaning on an arch of russet-colored stone, he looked about the chamber. "What does your sixth sense tell you about this place?"

"That we'll be okay."

A four foot-wide corridor stretched in front of them. They set off into it, as their boots scuffed on an uneven floor, inciting an unknown creature to scurry above their heads. Malcolm flicked his light upwards. The beam dissipated in unending darkness. "I wonder how high this goes?"

"Let's focus on what's in front of us." Under the distorted shadow of the lanterns, neither one had a desire to talk, nor whisper. The passage had few turns, although at times the walls closed on one another, forcing them to lean against the chilled limestone and slide sideways.

Time and distance were beyond reckoning. They cautiously descended a long incline which bottomed into a larger space with a flat floor. Water seeped across the floor in a shallow chute stained in a spectrum of red and orange. Jolene walked through the trickle and tracked imprints of her right sole that faded with each successive pace. Malcolm, uneasy with such plain evidence of their presence, circumspectly stepped over the flow. Glancing at the source of the water in the stone wall, he exclaimed, "I'll be damned."

She turned about and elevated her lamp. "What?"

Hieroglyphs of crude human and four-legged figures spanned

the wall, coarsely painted in brick red. Characters, perhaps of Asian origin, accented the figures. "What in the hell do you make of this?"

Jolene outlined one of the figures with her fingers. "I have no idea."

"Rather disturbing subjects and venue for a mural."

She scratched the figure. "Do you think its paint…or blood?"

Malcolm cast furtive glimpses above and behind. "Let's go."

After a brief gentle ascent, the passage narrowed and bent sharply to the right. They entered a chamber that was similar to the cave where they had begun their subterranean trek, only larger. The rain outside the entrance cast a dreary mood within the space. Nonetheless, Malcolm felt twenty pounds drop from his shoulders. Jolene shifted her attention to a corner of the cave and lifted a coil of copper tubing, matted with the green powder of corrosion. Beneath lay a pile of moldering canvas bags.

"Supplies for a still," she guessed.

Malcolm said, "I'm afraid to ask who's still this might be." His eyes caught sight of a six foot tall passage that opened into an alcove. Approaching the portal carefully, his torch shone about the perimeter of the cell. "Holy shit, we found King Solomon's mines."

Wooden crates full of corked whiskey bottles and ceramic jugs lay about the walls. Malcolm took a tentative step. "This can't be. There might be more than a hundred gallons here." He cautiously upheld a bottle and swirled the amber liquid within to appraise the clarity. He tugged the brittle cork from the neck of the bottle and sniffed the contents. He afforded himself a swig. Molten honey flowed down his esophagus and instantly warmed his core.

"Jeez, that's smooth. From the looks of the dust and style of these bottles, this stash has been aging here for years, probably decades."

"Malcolm, there's several open crates to your left. By the looks of the disturbed dust, it appears that someone else has shopped at this liquor store recently."

Malcolm rested the bottle back down in place and back pedaled into the main chamber. As they approached the egress from the cave, they came across a horizontal rock, the height and width of a small table. A pile of bones leaned lengthwise against one another to form a crude pyramid, underneath, within the small cavity, lay a military medal with a frayed ribbon. Several candles were scattered about on the flat surface.

Jolene asked, "What is this? It looks like some sort of shrine."

He replied, "I think I've had enough of cave dwelling."

They eased outside, rejuvenated by the fresh air, despite the mist and deep woods that suspended over them. They were at a lower elevation than the cave on the other side of the ridge spur, closer to the valley floor. A muddy tract lined with rain-laden plants led down the slope amid ancient hemlocks. The explorers looked at one another and set upon the viscous path.

Stones and gnarled roots grasping at their feet forced them to carefully choose their steps, while shiny rhododendron leaves brushed their faces with rough, wet swipes. Fifty paces from the cave, a spring gurgled through a mossy knoll and organized into a brook. The quick water trickled along the foot path which stretched out into a two-track wide enough for a vehicle.

"Fresh water for the mash and a place to park a truck with supplies," Jolene said.

The lane descended for another fifty paces until the water diverged away to their left and the track climbed a slight grade. They halted at the crest and peered through dull hemlock needles upon a clearing anchored by an imposing lodge, the MacDougall homestead.

Malcolm groaned. "I should have known the destination for this adventure wasn't a good one."

Neither dared to move further, sheltered in the deathly quiet behind the verdant curtain. Crows radiated from the far side of the barn, heralding their displeasure with the lurkers in the woods.

Jolene murmured. "I guess we need a plan."

"Great time to decide that."

CHAPTER XLIX

Jolene sized up the situation. "We should try to find a way to the far side of the farm."

Malcolm grumbled, "Bad idea, this place shouldn't be disturbed. The inhabitant's hospitality leans more to hostility."

"You want to go back and crawl through the tunnel?"

For Malcolm, both options held little appeal but trespassing through the MacDougall property would be a stroll in a city park compared to returning through the narrow tunnel. He pointed to the pasture. "I suppose if we got behind the barn then over to the far tree line, we could parallel the lane back to the road. We'll be under cover from the hedge row most of the way and no closer than fifty yards to the house." Malcolm mentally noted that was within range of buckshot—at least the black dog was no longer a problem.

"If we reach the road we can hitch a ride back into town. Even if we have to walk, it's only about four or five miles." His skepticism was palpable. She said, "That, or the cave."

"I guess your plan is the lesser of two evils…sort of."

They backtracked to where the brook crossed the path and walked away from the house, staying inside the tree line. Progress was slow as they picked through grabby brush and briars and navigated over numerous dead tree limbs. Malcolm's muscles flexed with increasing tension, his body feeling like a stretched rubber band, ready to snap if he heard a shout, or worse, gunshot. But the crows seemed to be the only ones aware of their presence.

To their right, they drew even with the brushy fence line that divided the back yard of the lodge from the overgrown pasture behind the barn. "Just stay on this side of the fence and below the level of the hedge."

"Really, Malcolm? I thought I'd rather knock on the back door and ask permission to look at their garden."

Their eyes engaged. He smiled. "Maybe while sipping some of that fine whiskey."

Vulnerability gnawed at their backs with their first steps out of the woods. They got into the pasture through a hole in the fence that only had a single strand of barbed wire laying on the ground. They dashed towards the bushes between the lodge and the pasture in a half crouch and followed the line with the manor on their right. With less than thirty paces to the barn, they confronted a gap in the hedge where only sporadic vines threaded the wire fence. The lack of sound was disquieting as they squatted and calculated the distance to the barn.

Malcolm stretched his neck around the hedge and came face to face with the sulking lodge, reflecting an atmosphere of neglect and decay. Heart thudding in his chest, Malcolm ventured into the unprotected space. Jolene's footsteps were close behind as they cut loose across the thick wet grass. Two crows jolted out of the far side of the barn and catapulted into the sky with highly-animated wings, decrying the trespassers. Malcolm instinctively grabbed Jolene and pulled her down into the sodden turf. They huddled together, the essence of damp earth rising into their nostrils.

Malcolm said, "Damn crows, shouldn't be so jittery." He perused the perimeter of the pasture and rolled his head to the right. He was eye-level with the clan MacDougall cemetery. A paralysis overcame his senses as he read the crudely gouged inscription on the nearest monument. "Jolene! The headstone, it's Archie MacDougall's! From six years ago!"

Jolene said, "That means that he's been…"

"Dead," Malcolm finished. "And not involved with the missing pets."

They looked at one another, reflecting on Archie's demise and adding to their sense of dread. Unconsciously, Malcolm glanced past the headstone and into the kitchen window where he had once dared to play the role of voyeur while looking for the black dog. An indistinct female form glided behind the panes, her lips moving as if speaking. Hair in a tight bun, she appeared to be wearing a high-collared, long dress. The figure halted and swiveled her head to gauge the visitors crouching by the fence. Jumping to his feet, Malcolm reeled away from the hedge and grabbed Jolene's hand. "Time to move!"

CHAPTER L

Scrambling from the graveyard, they reached the barn wall on the far side from the lodge. Malcolm leaned his back against the weathered boards, waiting for the inevitable uproar that signaled their discovery. But none came. He estimated the distance of the open space between the end of the barn wall to the. Two hundred feet, two hundred long, exposed feet.

Jolene edged a few steps ahead of him and peered into a gap between a pair of cockeyed slats. "We could crawl in here."

Malcolm consciously kept his voice low. "Inside the barn?"

"If we get to the side door on the other side, we'd have a shorter distance out in the open to the woods."

"What is it with you and confined spaces?"

She extended a leg through the breach. "At least no one can see us from the house in here."

Pinholes of anemic light dotted the chinks in the wallboards, providing enough illumination to discern free space from clutter. Moldy forages and machine grease mixed with dust in the stale air, prompting them to breathe minimally. They shuffled past four worn box stalls, bedded with straw that hadn't been stirred in a generation; on their right, rested a manure wagon with rotting boards and flat tires. An oil and paint-stained workbench bordered the far wall, decorated with a vise, and various springs, bolts, and drive belts. A variety of animal skulls collected dust on a shelf above the bench. Rusty forks, shovels, and scrapers hung from nails driven into the walls and posts. The only order amid all the entropy was a tidy stack of sliced white bread, separated by seams of brown paste. Mice or some other creatures had chewed the edges of the pile.

The prowlers picked their way through phlegmy cobwebs, which clung tenaciously to their damp hair and clothes. No barn swallows flew in this foul space. Tracking along a gradient of light, the far corner held promise for an exit. Malcolm tapped Jolene's shoulder. "Let's head for there and see what the view looks like."

They edged around the last box stall and discovered a portal wide enough for cattle. There was no door, but their path was obstructed from a mass on the ground. Acclimating his vision to the contrast in shadows, Malcolm blurted out, "Sweet Mother of God!"

Jolene gagged, her face contorted as she looked upon a body of a man, spread-eagled on his back. His torso lay in the obscurity of the barn, the jeans on his legs stained by muddy water across the threshold. The glistening cartilage of his trachea suspended from a gash across his throat.

Malcolm's ears pounded as he gently pushed Jolene. "Move, dear God in heaven, move!"

They rushed towards the woods. On reaching the wooden fence, Malcolm boosted Jolene over the top. He scrambled up the rails, his hands sliding on the uppermost rail, which was sticky with a thin wash of blood. His feet thumped onto the ground while crows in the trees above denounced their escape. "Damn birds," Malcolm cursed under his breath. The refuge of the woods within reach, they stopped three steps past the fence. A wraith lurched from under the misty canopy. The cadaverous figure lifted his head, staring with empty eyes. One hand grasped a rusted sickle, the other dangled limply. Sighting the intruders, the creature stumbled forward but slipped and fell grunting into the embrace of the wet soil. Slathered in fresh mud, the fiend somehow managed to regain its feet. Coughing and spitting, a voice wheezed, "You wontik h-a-m-m-m-m, you cat h-a-m-m-m-m!"

Malcolm pivoted. "Head for the lane, we can outrun him down the road." They retreated back over the fence, as they heard "Git oar m-m-m-m-m-t!" from behind. Malcolm nearly slipped and fell on the wet ground. Cutting across the weedy yard, they approached the forsaken Ford coupe at the split in the lane. The way was open to the road beyond, or so it seemed.

A man, attired in tweed trousers, long-sleeved white shirt, and a vest, loitered by the hood of the coupe; one foot on the ground, the other hoisted on the tarnished front bumper. Cold hatred snaked from a pair of yellow sparks within the shadows under the brim of a fedora. Stunned by the phantom, Malcolm and Jolene recoiled back towards the lodge. The filthy stalker had managed to clamber over the fence and shuffled towards them. From a rocking chair on the porch, a woman, dressed in a high-necked blouse and a long skirt, held a hand axe on her lap, humming a melody while casting a pleasant smile in their direction.

Blood drained from Malcolm's face. He panted. "Jolene, the white truck! See if you can start MacDougall's truck!" She sprinted to the Chevy. Malcolm scanned the ground and spied a five-foot piece of locust fence board. He clutched the scrap from the sucking soil only to have the rotted timber break in half. "Son of a bitch!" He cast the spongy artifact in a two-handed hammer throw to check the advance of the zombie-like creature, missing wildly. His intended target threw back his head and howled.

Jolene called out from behind the wheel of the truck, "Malcolm, the keys are here!"

Malcolm ran to join her, yanked open the passenger door and tracked the creature's progress. Listing to one side, it closed in on them at a slow pace, blood-stained sickle still in hand. As it got to within thirty feet of the truck, the blood shot eyes were the only discernible feature of the face beneath the filth, a thin dribble of spittle draining from the corner of his mouth. The creature piloted its rambling gait towards Malcolm.

Jamming a foot on the clutch, Jolene turned the ignition, only to hear an unproductive grind. "Shit, come on," she pleaded. Her hand tried the starter again. She paused. "Damn it, I should have known, it's got a manual choke!"

Malcolm looked about the bed of the truck; a tire iron lay next to the wheel well. Clenching the cold steel in his hand, he turned to face his attacker. The fiend swiped with the blade. The sickle whirred through empty air as Malcolm juked away with a sidestep.

Halting to check its bearings, the grimy demon turned to stare at Jolene lurch through the windshield. With a hideous smile and choking hiss, it made for the driver's side of the truck, sickle raised overhead.

"Jolene, shut the door!"

Yellow teeth flashed as the sickle clanged off the side mirror. Jolene recoiled away from the window, falling back across the seat. The revolting face popped through the window, slavering and trying to speak with a protruding tongue. "Ar-r-gy ho isha isha cra-a-a-w."

Jolene bent a knee over her chest and lashed out with the heel of her boot. The monster flinched as his head spun from the blow on his jaw. Spitting out a tooth and sputtering blood, it vented cruel humor with a laugh. "Ha-augh shawpa wijawa."

Shouting to distract the attacker, Malcolm bounded around the hood of the truck. The figure stumbled forward to engage him, sickle raised. Malcolm gripped the tire iron in both hands to parry the blow. With a sharp metallic ring, the blade skimmed off the iron and slashed across Malcolm's upper arm, seeping blood into his jacket. The beast weakly grabbed Malcolm's sleeve with his free hand and stretched its neck, mouth open as if to bite.

Digging his feet into the slippery ground, Malcolm steadied his balance, spun his waist, and swung a furious blow with the iron. Screaming, the goon released his hold and staggered back several steps. The sickle-loaded hand wavered erratically. To Malcolm's dismay, the feral eyes once again targeted Jolene, drawn by the noise of the defunct ignition. Unaware of his approach, she was focused on starting the truck.

Malcolm lunged at the attacker but lost his footing in the mud. He shouted, "Jolene!" She looked up with wide eyes to see the sickle hovering over the window.

A percussive shock roared through the clearing and echoed against the lodge and surrounding forest. The hostile eyes of the beast faded into a glassy stare. Nonchalantly dropping the blade, it slumped onto unsteady knees, gargled red foam and collapsed face first into the soil.

Pushing up off his hands and knees, Malcolm gaped at the crumpled body by the front tire. He faced Jolene. She forced a thin smile. Malcolm looked beyond the truck bed; seventy-five yards away sat a black and white Ford, Chief Cal and Tommy standing behind open doors. Tommy held a rifle with a scope. For a brief moment, Malcolm and Cal beheld each other. Malcolm let the iron drop to help Jolene out of the truck. Giving her a hug, his voice unsteady, he asked, "You okay?"

"I'm alright."

Moments later, the police cruiser pulled up to the truck. Cal pushed himself out of the passenger seat. "Kind of odd that we should shoot a man for trying to keep his truck from being stolen."

Malcolm looked upon the lifeless form once again. "What man?"

"John MacDougall himself, I imagine."

Malcolm and Jolene shared a look of disbelief. Jolene pointed to the barn. "There's a body, over in the doorway to the barn."

Cal suggested, "Tommy, go take a look-see for me, please." Sizing up the disarrayed couple, Cal added, "I'm, afraid it's probably John's son, Lester MacDougall."

Jolene gasped, "I knew him from school."

Cal shook his head. "The tragic consequences of hiding dark family secrets."

Malcolm surveyed his surroundings, paying particular attention to the Ford coupe and front porch, both of which were unattended. "There were others here, a man by the car and a woman on the porch."

Cal scanned the yard. "Others, Doc? We didn't see anyone else when we arrived."

Malcolm looked at Jolene, who remained silent. Compressing a hand on his cut arm, Malcolm said, "We found a gravestone for Archie MacDougall, he's been dead for six years!"

"Doc, I suggested to you some time ago not to believe all the gossip around here, saturates this valley like sticky glue. The more you get involved, the more you get entangled. It's kind of gotten to be second nature for you. I did a little research this past week. Archie MacDougall died of a heroin overdose in Philadelphia."

"How did John MacDougall keep that a secret?"

"Happened in a flop house, in a big city, far from the valley. Not the kind of thing a family would want in the local obituaries."

Jolene queried, "How did you know we were here?"

"I got word of your expedition with all of the commotion with the professor going off in the ambulance. I cornered the Calhoun boys as they were trying to take your truck and found out about the cave adventure. When we got up there and you weren't to be seen, I considered the next place you might have wandered into if you crawled through the cave."

Jolene's eyes popped. "You knew of the cave?"

Cal chuckled, "Wouldn't even think of going through myself, but like any good moonshiners, the MacDougalls had an escape hatch from the bootlegging days. I figured arriving at the far end of the cave at the wrong time and place could lead to unpleasant things." Cal examined Malcolm. "You seem to like this place, Doc."

Malcolm felt ill at ease with that proposition. "Hardly."

"Chief Cal," Tommy called out, "I think this here body is Lester MacDougall."

Cal winced. "I guess we'll need to arrange two coroner's inquiries."

Malcolm asked, "Should you have MacDougall's corpse tested for rabies?"

"That's up to the coroner, but I'll make a recommendation. I think it's time we get you two where you belong, which isn't here. Better make a call on the radio." Limping to the car, Cal said, "Doc, you should take up fishing, builds patience, which happens to be a virtue."

*****

Waiting for the ride home, Malcolm leaned his head against the rear window of the police car, surrendering to the exhaustion and damp chill in his body. Emergency personnel and state police tramped about the homestead. Someone had bandaged his arm, he couldn't remember who. "For the sake of my sanity, tell me you saw that guy in the hat by the old Ford, and the woman on the porch, too."

Jolene leaned against his shoulder. "I'd rather not."

His heavy eyes gathered in a last view of the lodge, so much madness and strife conjured up from one place. The clouds had dispersed into a patchy cover and soft evening light brushed the masonry of the stone chimney nearest to him. He started up in his seat. Uncertain if he witnessed the first one, a second and third bat squeezed out of a seam under the roof. "Jolene, you aren't going to believe this…"

CHAPTER LI

For the first time in a year, Malcolm sensed life in the valley was returning to normal. The corn was poking out of the ground, the wheat was sprouting heads and the first-cutting of hay was ready to mow. Placid cows ambled about hillsides, white and black flotsam and jetsam in green seas of grass. Malcolm wheeled into Jonas P. Yoder's barnyard, needing eggs for a weekend omelet he had promised Jolene.

He stepped from Lucille, welcomed by Fritz who barked and ran to the vehicle, his ungainly younger companion behind him. "You've grown you goofy thing." He vigorously rubbed the back of the younger dog. Rachel ambled from within the milk house.

"I see you have my dozen eggs, Rachel."

A skittish barn cat bolted in front of the maiden. Taken aback, she lost her balance and dropped the carton of eggs. The puppy bounded over to appraise the spilled booty. Rachel shrieked, "*Geh weg!*" and kicked at the hapless dog.

"What's the matter, Rachel? Are you okay?"

She glared at Malcolm with querulous eyes as she backstepped, "*Geh weg!*" Spinning about, she jogged around the corner of the milk house.

Stunned, Malcolm turned to find Jonas P. standing next to him. "Jonas, is Rachel okay?"

The farmer shook his head. "Something is wrong, she isn't herself. Doesn't read or look at us. We tried some hexing with Amos M. Yoder, but it's not doing any good."

Malcolm's pulse quickened. "Has she seen a doctor, Jonas?"

There was no reply. *"Jonas!* Has she seen a doctor?"

"Nein, Tierarzt. Kein Doktor."

"Jonas, please get Rachel and your wife, I'm driving you to the hospital in town."

"That is kind, but we don't need the help, *Tierarzt.*"

"Please Jonas, let me take you in. For God's sake, let me take you."

Tears brimmed over the eyelids and onto the farmer's weathered face. From then on, he disavowed his belief in a brutal God that demanded that he live under the crushing rule of religious tyranny.

*****

Malcolm carried plates and silverware and placed them on the little patio table by the edge of the garden. The plates were laden with omelet. Sadly made with white eggs from the grocery, he thought. Peonies and iris were playing their part in the garden canvas to celebrate the warm morning. Jolene sipped her coffee. He kissed her cheek.

She reached her hand up, stroked his hair. "So Chief Cal thinks that MacDougall had rabies?"

"Yes, but the final autopsy report hasn't been made public."

"Certainly would explain his overly belligerent behavior."

He sighed, "And I suppose the missing pets in the area. It seems odd, being that far gone and still able to consciously stalk and kidnap folk's pets. The distance would have required a vehicle."

"Well, who else could it have been?"

"It sure wasn't his brother Archie."

Jolene considered a likely possibility— perhaps someone who wanted to create the illusion it was Archie. She asked, "Are you going to let the university experts know that rabid bats might have been roosting in the MacDougall home all this while?"

"I already called Lois to get in touch. I understand they'll return when Sinclair's leg has healed."

Jolene smiled and looked into the distance. "I wonder if they'll receive a warm reception from some of the family elders?"

Malcolm lowered his plate on the ground for Precious to finish. "I'm heartbroken for Jonas P., his family—and Rachel."

"Any news from the hospital?"

He shook his head.

"Do you think she has rabies?"

Sitting back, he viewed silver clouds drifting over the ridge. "It doesn't look good."

"Malcolm, you tried. The whole rabies problem, you tried. That's all you can do."

"I could have done more."

No, Malcolm. There are some things you can't overcome in this valley, suspicion of outsiders, and embracing religious dogma over science are but two of them."

"Indeed." The irises fluttered in a light breeze as he mindlessly swirled his coffee mug. He grunted. "I still can't believe how I got pulled into all of this."

"This is Galloway County." She pointed to ridges that slanted away into the haze on the horizon. "Malcolm, look at these hills around you. They're a trap for those who live here. A beautiful trap, but a trap that will ensnare people into isolation and ignorance, and where there are no secrets."

"Except the fate of Archie MacDougall and the cache of moonshine up in the cave."

"Would you like to go plunder that?"

"Not in this lifetime."

"How about sing a duet with the lady on the porch?"

"Stop!"

"What? A science guy like you? Scared of ghosts?"

Malcolm offered Precious a piece of toast as he spoke softly, "You're such a food hog."

Lying on the patio, Precious relaxed, closed her eyelids and wagged her tail.

EPILOGUE

In early summer, there was a funeral at the farm of Jonas P. Yoder and a wedding at the farm of John Y. Peachey. Both were unusual events for the valley—the funeral because of the young age of the woman who passed away, the wedding because it was out of season and none of the bride's family attended.

Later that summer, Malcolm conspired with John Y. to organize a rabies vaccination clinic. Yellow topped buggies crowded onto the farm, one large table was assigned for canine immunizations, while others were piled with baked goods. Malcolm engorged moon pies to be worthy of a draft horse's sufficiency.

As the clinic wound down and evening milking approached, the sun descended once again towards the embrace of the timeless ridges. Rays of light stroked the valley, emblazoning scripture on a billboard at the Nahum Y. Peachey farm. *'Behold, the Name of the Lord cometh from afar, burning with his anger and dense clouds of smoke; his lips are full of wrath, and his tongue as a devouring fire; Isaiah 30:27.'*

Farther down the valley, the dwindling sunlight glowed about a solitary figure. Nahum Y. Peachey hunkered in his black-topped buggy, parked on the road at the entrance to John Y. Peachey's lane. He thought his father's use of the billboard on his farm. A tool used to batter passersby with judgement and to reflect on their burden of sin. He sighed. Sizing up the barnyard clinic, he weighed his choices and burdens of sin. His collie patiently sat by his side on the uncomfortable bench seat.

A note regarding Amish characters in this book. The characterizations of members from this community in this novel are based on personal observations, experiences, and conjecture to enrich the story. At times, in the course of normal interactions, particularly for practitioners who develop a working relationship with their Amish clients, German might be spoken in a conversational manner, such as greetings, simple questions regarding their animal patients, etc. It is understood that the dialect of Pennsylvania German spoken by the Amish is decidedly different in grammar and pronunciation than that of a native German. However, deciphering this dialect into the written word can be subject to regional preferences, accents, inclusion of English, and vernacular terms. Thus, a more pervasive German (as spoken in Europe) was used for ease of reader clarity, although given the simplicity of the conversations in this book where German was inserted, the written dialects are often identical. I am indebted to Dr. Ellen Weigert for her help in translation and ensuring that the less formal use of German was applied in this writing.

Bibliography

Constantine DG. Rabies transmission by air in bat caves. Public Health Service Publication No. 1617. Atlanta, GA: US Department of Health, Education, and Welfare, Public Health Service; 1967.

Gibbons RV. Cryptogenic Rabies, Bats, and the Question of Aerosol Transmission. *Annals of Emergency Medicine*. 39:528-536, 2002.